DC/AC Converters

DC/AC Converters

Analysis, Modeling and Design Consideration

Nikolay Hinov

MDPI

Basel • Beijing • Wuhan • Barcelona • Belgrade • Novi Sad • Cluj • Manchester

Author
Nikolay Hinov
Department of Computer Systems
Faculty of Computer Systems
and Technologies
Technical University of Sofia
1000 Sofia, Republic of Bulgaria

Editorial Office
MDPI AG
Grosspeteranlage 5
4052 Basel, Switzerland

For citation purposes, cite as indicated below:

Lastname, A.A.; Lastname, B.B.; Lastname, C.C. *Book Title*. Series Title (optional). Year, Page Range.

ISBN 978-3-7258-4058-8 (Hbk)
ISBN 978-3-7258-4057-1 (PDF)
https://doi.org/10.3390/books978-3-7258-4057-1

This research was funded by the Bulgarian National Scientific Fund, grant number КП-06-Н57/7/16.11.2021, and the BPC was funded by КП-06-Н57/7/16.11.2021.

Dedication

This book is dedicated to the 50th anniversary of the establishment of the Department of Power Electronics at the Technical University of Sofia.

With deep respect and gratitude to all lecturers, researchers, doctoral candidates and students who, during these five decades, have built traditions, inspired innovations and contributed to the establishment of the Department as a leading school in the field of power electronics in Bulgaria and abroad.

The Department of Power Electronics is not just a scientific center—it is a spiritual home of thinking, searching, and people dedicated to the development of power electronics. Here, knowledge is not only taught, but also created; ideas are not only discussed, but also born and implemented; young talents are not only trained, but also inspired to change the future.

This book is a modest contribution to the mission of the Department—to create knowledge, train specialists and develop science in the service of society.

I sincerely wish that the department will continue to be a source of ideas, innovations and inspiration for future generations of engineers and scientists.

Contents

List of Figures and Tables

Abbreviations

AC	alternating current
B	load factor of the parallel CSI
BCM	boundary conduction mode
$cos\varphi_L$	power factor
CCM	continuous conduction mode
CSI	current-source inverter
DC	direct current
DC/AC	converters and inverters
DCM	discontinuous conduction mode
EMI	electromagnetic interference
$F(\omega t)$	switching function
f	control frequency
$f_{(1)}$	frequency of the reference sine wave
f_S	frequency of the sawtooth voltage
GTO	double-operation thyristor
HCM	hard commutation mode
HVDC	high-voltage direct current
IGBT	insulated gate bipolar transistor
K_P	ripple coefficient of the current in the AC circuit
k	coefficient of oscillation
k_A	aperiodicity coefficient
M	magnetic coupling coefficient
M_A	ratio of the amplitudes of the control and sawtooth signals
M_f	ratio of the frequencies of the reference sine wave and the sawtooth voltage
n	number of the harmonic
NCM	natural commutation mode
ODE	ordinary differential equations
P	active power in the load
PRI	parallel resonant inverter
PWM	pulse width modulation
Q	quality factor
RI	resonant inverters
RIED	resonant inverter with energy dosing
RIRD	resonant inverter with reverse diodes
SPWM	sinusoidal pulse-width modulation
SPRI	series–parallel resonant inverter
SRI	series resonant inverter
TF	the transfer function
t_{qc}	thyristor turn-off time
U_{Contm}	amplitude of the sawtooth signal
U_d	DC power source
U_{Trm}	amplitude of the control signal
UPS	uninterruptible power supply
VSI	voltage-source inverters
ZCS	zero current switching
ZVS	zero voltage switching

β	detuning angle of the parallel load circuit
δ	damping of the resonant circuit
η	efficiency of the DC/AC converter
$\varphi_{(1)}$	the phase difference in the first harmonic of the current in the AC circuit
ν	frequency coefficient
ν_A	quasi-frequency coefficient
ω	control frequency
ω_0	natural (resonant) frequency

About the Author

Nikolay Lyuboslavov Hinov is a Professor (Associate) in the Department of Computer Systems at the Faculty of Computer Systems and Technologies at the Technical University of Sofia, Bulgaria. Born in 1970 in Lukovit, he graduated with a master's degree in "Electronics and Microelectronics" at the Technical University of Sofia in 1995 and defended his PhD dissertation in 1998 in the field of high-power electrical energy converters with industrial application. In 2024, he also defended his dissertation for the degree of "Doctor of Science", dedicated to the model-based design of power electronic devices with the application of artificial intelligence techniques.

His professional career is inextricably linked to the Technical University of Sofia, where he held various academic positions—from senior assistant to professor. Over the years, he has held leadership positions, including Deputy Dean of the Faculty of Electronic Engineering and Technology (2011–2019) and Head of the Department of Power Electronics (2019–2025), as well as Secretary of the Bulgaria Section CAS/SSC Chapter of IEEE.

His scientific interests include industrial and power electronics, artificial intelligence systems, electronic converters, autonomous and electric vehicles, smart grids and cities, as well as modeling and optimization of mechatronic systems. He has over 300 scientific publications indexed in Scopus and Web of Science, and over 800 peer-reviewed reviews in IF and/or SJR journals, and he has participated in more than 50 research projects, leading 10 of them. He holds 10 patents and 5 utility models, and under his supervision seven doctoral students have successfully defended their dissertations. He has initiated, organized and edited more than 10 special issues in various IF journals. He has participated in the organization of over 15 international scientific conferences indexed in Scopus and/or Web of Science.

He is the author of textbooks, laboratory manuals and one monograph indexed in Scopus. He is a member of the editorial boards of three international scientific journals with an impact factor. He has received a number of awards, including the title of Senior Member of IEEE, the Gold Badge and the Gold Medal "Prof. Assen Zlatarov" of the Federation of Scientific and Technical Unions in Bulgaria, an award for the best project of the National Science Fund (Bulgaria) in the field of Technical Sciences, reported in 2024.

Foreword

It is great pleasure and honor to write the foreword and introduction to this monograph by Professor Nikolay Hinov, on the theoretical foundations, modeling techniques and practical design considerations of DC/AC converters in modern electrical energy conversion—it offers an in-depth exploration of the emerging power electronics technology.

Power electronics has become a cornerstone of new sustainable energy systems (the general electrification of the society), electric transportation, industrial automation and energy savings. At the core of this transformation are reliable and efficient power converters. This book delivers a nice, comprehensive and structured discussion of DC/AC converter topologies, with particular emphasis on current-source, voltage-source, and resonant inverter configurations.

This work is distinguished from many other sources due to its unified analytical methodology. By focusing on the electromagnetic processes within inverter circuits, Professor Hinov effectively bridges the gap between theoretical analysis and engineering practice. This approach fosters consistency in modeling and facilitates comparative design across various power converter types—an invaluable asset for researchers and practicing engineers.

The monograph stands out not only for its technical merits but also for its pedagogical approach. Its treatment of both steady-state and transient phenomena, supported by models implemented in MATLAB/Simulink and LTspice, gives the reader practical tools for simulation, modeling, and system design.

Professor Hinov brings to this work over three decades of outstanding academic and professional experience. His substantial contributions to modeling methodologies and to the promotion of systematic design practices have left a lasting mark on engineering education and research in Bulgaria and internationally.

What particularly impressed me is the way Professor Hinov combines analytical models with a profound understanding of real-world engineering constraints which are important in order to master the power electronics technology. With more than 300 publications, leadership in international projects, and a longstanding record in teaching and mentorship, he brings an authoritative and insightful voice to this important field.

This book is also a tribute to the academic excellence of the Bulgarian school of power electronics. Its dedication to the 50th anniversary of the Department of Power Electronics at the Technical University underscores a long tradition of innovation and distinction in the field.

This monograph will become a key reference for graduate students, researchers, and practicing engineers. I am confident it will inspire further advancements in high-efficiency and application-specific DC/AC converter systems, particularly in an era where digitalization and intelligent control are reshaping the future of power electronics.

I congratulate Professor Hinov on this significant achievement and warmly recommend this book to all seeking a thorough yet practical understanding of DC/AC converter systems. Enjoy the reading.

Frede Blaabjerg
Professor, IEEE Fellow
Aalborg University, Denmark

Preface

The motivation behind this book arises from the growing importance of DC/AC converters in modern power electronics. These systems are at the heart of technological transformation across energy generation, industrial automation, electric transportation and renewable integration. In recent decades, their role has become increasingly central to the advancement of efficient, reliable and flexible power processing.

The intent of the author is to provide a comprehensive, yet accessible resource that bridges the gap between theoretical foundations and practical implementation. While numerous academic papers and textbooks explore aspects of inverter technology, there remains a need for a structured work that unites detailed analysis, modeling approaches and design considerations in one coherent framework.

This book presents an in-depth study of DC/AC converter systems, focusing particularly on basic inverter topologies. The research advances a unified methodology for analyzing various inverter types by examining the electromagnetic processes in their power circuits. Mathematical models are developed for current-source, voltage-source and resonant DC/AC converters, supported by diagrams, operating waveforms and analytic derivations. Both steady-state and dynamic behaviors are discussed in the context of real-world applications.

The work is structured around analytical methods validated through established engineering principles. No previously published material requiring third-party permission has been reused in this volume; all models, derivations, diagrams and simulations have been independently developed by the author for the purpose of this monograph.

This book is the culmination of over 30 years of teaching, research and professional experience in the field of power electronics, and it is intended to serve as a reference and inspiration for students, engineers and researchers striving to contribute to a more efficient and sustainable energy future.

Nikolay Hinov
Author

Acknowledgments

I would like to express my sincere gratitude to my professors and teachers from the Department of Power Electronics at the Technical University of Sofia for their professionalism, dedication and inspiration. Thank you for patiently and dedicatedly passing on your knowledge, stimulating my critical thinking and showing me the importance of precision and responsibility in engineering work. Your support and the high standards you set were a leading factor in my development and motivated me to confidently move forward in my professional path.

I express my deep gratitude to Prof. Nacho Nachev—founder of the Department of Power Electronics, for his remarkable contribution to the development of academic and research activities in this field. Thanks to his vision, leadership and tireless work, a sustainable foundation was created for the training of highly qualified engineers and for the establishment of the Department as a leading scientific school in the field of power electronics. His contribution leaves a lasting mark not only in the history of the department, but also in the future of all who have the privilege of studying and working here.

I express my sincere gratitude to Prof. Nikola Gradinarov for his invaluable guidance, support and inspiration during my work. Under his expert guidance, I had the opportunity to upgrade my knowledge and skills, develop my analytical thinking and gain valuable research experience. Thank you for your patience, high demands and invaluable advice that encouraged me to achieve more than I myself believed was possible. Working with Prof. Gradinarov was an honor and a huge professional inspiration for me.

I respectfully acknowledge the contribution of Prof. Evgenii Popov—the founder of modeling and automated design of power electronic devices in Bulgaria. With his pioneering work, scientific insight and dedication, Prof. Popov laid the foundations of a key direction in power electronics, which today is an integral part of engineering practice and scientific research in our country. Thank you for the inspiration, example and high professional standards, which continue to be a guide for all of us who have linked our development with this field.

My sincere appreciation is extended to Prof. Stefan Tabakov for his significant contribution to the establishment and development of the Department of Power Electronics throughout most of its 50-year history. With dedication, consistency and high scientific level, he contributed to the construction of stable foundations in education and scientific research in the field of power electronics. Thank you for the inspiration, long-term work and example of professionalism, which continues to be a reference point for generations of young engineers and scientists.

I sincerely hope that the young colleagues who are now joining the Department will preserve its spirit, uphold its traditions and continue to develop its academic and scientific excellence with the same dedication and vision that shaped its distinguished history.

1. Introduction

Inverting is the process of converting direct current (DC) electrical energy into alternating current (AC). Devices that implement inversion through the means of electronics are called DC/AC converters (inverters). The areas of application of inverters are very diverse, the main ones being the following [1–4]:

- Power supply of electric drive systems, including electric vehicles;
- Power supply with high-frequency energy of various electronic technological devices and systems for the needs of: induction heating, ultrasonic processing, dielectric heating, laser technology, crystal growth for the purposes of semiconductor production, etc.;
- Power supply of alternating current consumers in cases where the only sources of electrical energy are direct current—for the needs of ships, airplanes, cars, etc.;
- Backup power supply of alternating current consumers in hospitals, banks, computer centers and data centers, communication systems, power plants, nuclear reactors, etc.;
- Power supply of remote areas or connection of power systems through high-voltage direct current (HVDC) power transmission lines (with low power losses) and its subsequent conversion into alternating current;
- In distributed generation systems of electricity with renewable sources such as: wind generators, photovoltaics, fuel cells, generators driven by the energy of sea waves, tides, small rivers, waterfalls, streams, etc.

Due to the need to ensure maximum efficiency at a minimum set power of individual elements and nodes of the DC/AC converters, various controllable-power semiconductor elements operating in a key mode are used for their implementation, such as single-operation and double-operation (GTO) thyristors, reverse-conduction thyristors, powerful bipolar transistors, MOS transistors and insulated gate bipolar transistors (IGBTs).

The formation of alternating current and voltage at the DC/AC converter output is achieved by alternating the on and off states of the semiconductor elements in the inverter circuit. The need to use controllable-power semiconductor elements in the energy conversion process is determined by the principle of operation of inverters, since during a significant part of the time interval, when the semiconductor element is turned off (closed), a positive voltage is applied to it. When using incompletely controlled power semiconductor elements (single-operation thyristors), their shutdown and restoration of their ability to be controlled is ensured by the operation of the power circuit or by additional switching circuits consisting of capacitors, inductors, uncontrolled or controllable-power semiconductor elements, etc. Switching elements can be common to all thyristors in the inverter, common to a group of thyristors, common to thyristors in a given phase or arm, or be individual for each thyristor [5,6].

1.1. *Classification of DC/AC Converters*

Depending on the method of current switching between the power semiconductor elements, DC/AC converters are divided into two main groups: dependent (grid) and independent (autonomous). In grid DC/AC converters (the simplified block diagram is given in Figure 1), the consumer of the power received from the inverter is the AC network. It also includes other sources of AC electrical energy (marked together with a generator) with a power much greater than the power of the dependent inverter, as well as energy consumers (Z_{L1}, Z_{L2}) [7–9]. The current switching from one power semiconductor element to another is performed by the voltage of the network (in the case of thyristor variants) or synchronized with it (transistor ones), and thus the frequency of the current in the AC circuit of the inverter is equal to that of the network. Depending on the type of AC grid, dependent DC/AC converters are single-phase and three-phase. The power circuits of dependent thyristor inverters coincide with those of the corresponding controlled rectifiers, with the exception of asymmetrical (incompletely controlled) rectifiers and those that contain a reverse diode at their output [10,11].

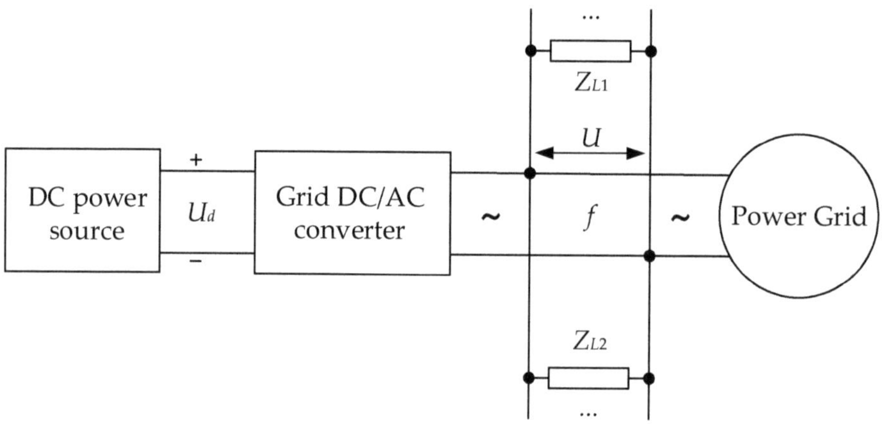

Figure 1. Block diagram of a single-phase grid (dependent) DC/AC converter. Source: Figure by author.

In independent (autonomous) DC/AC converters (with a block diagram as given in Figure 2) the frequency, shape and parameters of the output voltage and current depend on the following: the type of circuit and the load with their parameters and the control system. Therefore, no alternating current electrical energy is supplied to the output of these inverters from sources other than that generated by the autonomous inverter itself. The object of research in this book will be autonomous DC/AC converters and further in the manuscript the adjective 'autonomous' will not be used [6,12].

Depending on the nature of the electromagnetic processes developing in the power circuit, inverters are divided into three main groups: current-source DC/AC converters, voltage-source DC/AC converters and resonant DC/AC converters [4,13,14].

Current-source inverters (CSIs) are characterized by the fact that their output current (current in the AC circuit) i_{AC} has a rectangular shape with an amplitude equal to the value of the input current consumed by the direct current (DC) power source I_d (in the bridge circuit, and in other circuit variants with a corresponding proportionality coefficient). The shape, parameters and phase angle of the output voltage u depend on the type and parameters of the load [1,6,13,15]. Figure 3 shows the forms of i_{AC} and u when working with an active-inductive load and a resonant circuit.

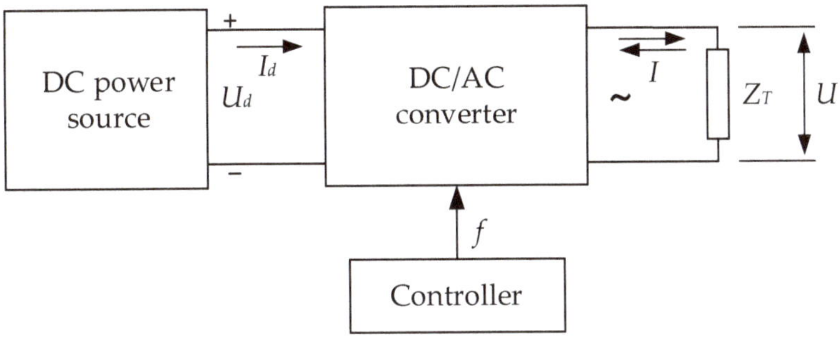

Figure 2. Block diagram of a single-phase independent (autonomous) DC/AC converter. Source: Figure by author.

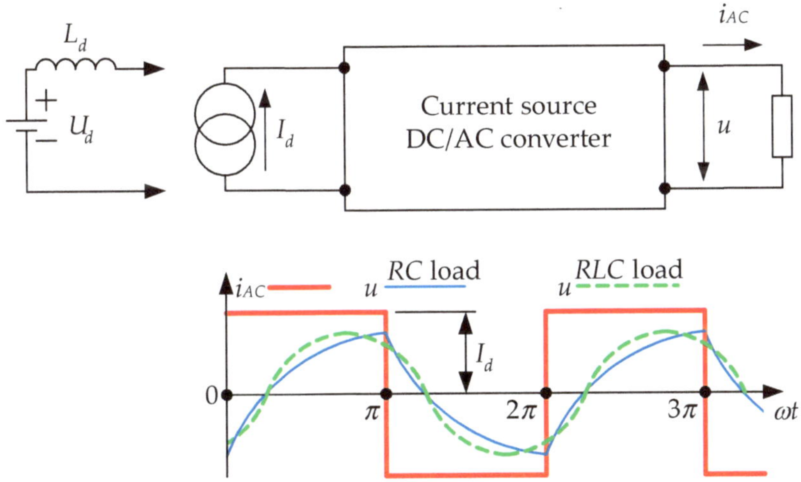

Figure 3. Basic structure of a current source-inverter (CSI) and time diagrams of output voltage and current. Source: Figure by author.

Thus, the DC source supplying the DC/AC converter operates in the current-source generator mode, which is most often implemented by connecting the voltage-source in series in the input circuit (by rectifying the mains voltage) and a choke with a very large inductance L_d. In a CSI equipped with thyristors, the commutation process is provided by capacitors, which can be connected to the load in different ways: in parallel (parallel DC/AC converters), in series (series DC/AC converters) or according to a complex scheme of the output circuit (series–parallel, parallel–series) (with capacitive increase), from which the corresponding names of the inverters arise [3,5,16]. It is characteristic of CSIs that they are operable in a relatively narrow range of load changes and have a strongly pronounced dependence of the output voltage on the load impedance; i.e., they have a soft external characteristic. There are single-phase and three-phase CSIs. Single-phase DC/AC converters are implemented in a full-bridge circuit or in a circuit with a midpoint of the output transformer [17].

In voltage source-inverters (VSIs), a periodic switching with alternative polarity of the voltage of the DC power source U_d (for full-bridge circuits, while in other circuit variants it is a multiple of this value) is performed to the load at the output [1,2,6,13]. The output voltage u is rectangular in shape, and the shape, parameters and phase shift of the output (load) current i depend on the type and parameters of the load (as shown in Figure 4). Thus, according to the laws of electrical engineering, the CSI and VSI circuits are dual, and one can unambiguously switch qualitatively and quantitatively from one to the other.

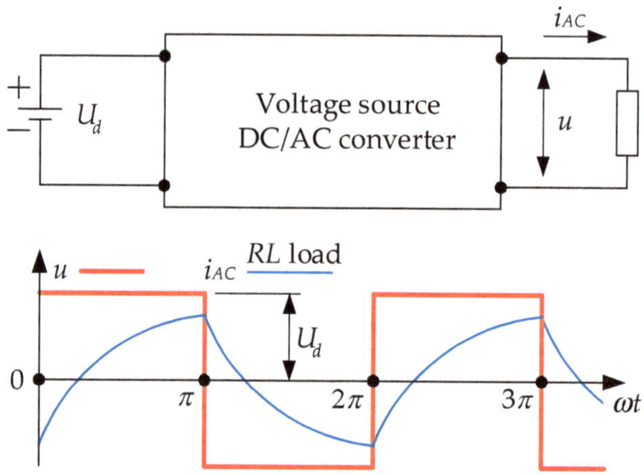

Figure 4. Basic structure of a voltage-source inverter (VSI) and timing diagrams of output voltage and current. Source: Figure by author.

VSIs are operable over a wide range of load changes and have a rigid external characteristic (weakly pronounced dependence of the output voltage on the load current). VSIs are also single-phase and three-phase. They can be implemented with fully controlled power semiconductor elements or non-fully controlled elements (thyristors) in a full-bridge, half-bridge or center-tapped transformer (push–pull) circuit.

In resonant inverters (RIs) presented in Figure 5, the load and reactive elements of the inverter circuit (resonant inductance L and capacitor C) form a resonant circuit at the output of the device. In the most common case of a series resonant circuit, the condition for the oscillatory nature of the processes in the power circuit is $R < 2\sqrt{\frac{L}{C}}$ [1,2,5,13,18]. In this case, the current through the load and the power semiconductor elements has a shape close to the sinusoidal one. For the case of a thyristor resonant inverter, the most common current shape is shown in Figure 5, and the power semiconductor elements are turned off due to the smooth decrease to zero in the current through them, or this is the so-called natural or soft commutation (zero-current switching). The shown current actually flows through the oscillatory loop every half-cycle. The capacitor participating in the resonant circuit is most often connected in series with the other elements, but there are also circuits with a parallel or series–parallel connection. Due to the lower switching power losses in resonant inverters, it is advisable to use them at increased frequencies of the output voltage and through them the highest frequency inverter circuits are implemented. To stabilize the operation of the inverter when the load changes, resonant inverter circuits with reverse diodes are used, and due to the design features of transistors, they are also necessary for the implementation of transistor resonant inverters. If the inverter's power source has a voltage-source characteristic, the inverter has an open input, and if it has a current-source characteristic, a closed input [16,17].

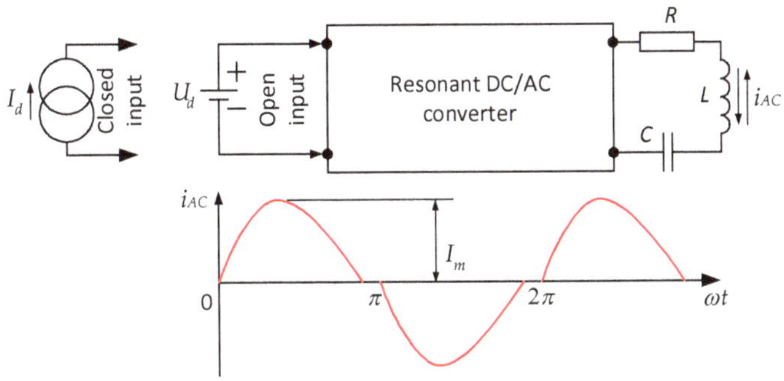

Figure 5. Basic structure of a resonant inverter (RI) and current shape in the AC circuit. Source: Figure by author.

1.2. Motivation for Writing and a Brief Description of This Book

The rapid development of power electronics has led to a growing demand for efficient, reliable, and optimized DC/AC converters, which are the foundation of modern society and industry. These converters play a key role in a variety of systems and applications. However, the design and analysis of DC/AC converters requires a thorough understanding of their operating principles, control strategies, and design tradeoffs.

This book is motivated by the need for a comprehensive and accessible resource that bridges the gap between the theoretical foundations and practical implementation of DC/AC converters. Although numerous research papers and textbooks exist, a structured, application-oriented approach that systematically covers modeling, analysis, and design considerations is lacking.

With this book, we aim to provide the following:

- A clear and structured analysis of DC/AC conversion techniques, including voltage- and current-source inverters and resonant topologies.
- Mathematical models that help engineers and researchers understand converter dynamics, performance constraints and stability considerations.
- Design guidelines and good practices for component selection, taking into account the specificities and operating modes of power electronic devices and systems).
- Case studies and simulations that illustrate and validate the proposed design methodologies.

By combining theoretical foundations with practical knowledge, this book aims to be a valuable reference for students, researchers, and engineers working in the field of power and industrial electronics. We hope that this work will contribute to the further development of highly efficient and energy-saving DC/AC converters that will shape our sustainable future.

Brief summary of the book's contents.

The first chapter of the book presents the basics of the process of inverting electrical energy, defining the concept of an inverter and classifying the different types of inverters depending on their operating principle and application. Special emphasis is placed on the distinction between dependent (grid) and independent (autonomous) DC/AC converters, with the attention throughout the book being focused mainly on autonomous solutions. The introduction also justifies the motivation for the creation of this work, related to the need for a comprehensive and practically oriented source in the field of power electronics.

The second chapter is devoted to an analysis of the different types of DC/AC converters. First, current-source inverters are examined, emphasizing their principle of operation, basic circuits, behavior with active and active-inductive loads, and the characteristic features of their external and input characteristics. Then, attention is directed to voltage-source inverters, including single-phase and three-phase configurations, bridge and half-bridge solutions, describing various methods for forming and regulating the output voltage through pulse-width modulation. Approaches for eliminating harmonic components are also considered. The subsection on resonant inverters explains the principles of resonant commutation, the topologies of various types of resonant circuits and the conditions for their operation with minimal switching losses. The last part of the chapter introduces a unified approach to the analysis of DC/AC converters, based on the study of electromagnetic processes in series RLC circuits.

The third chapter is devoted to the modeling of DC/AC converters with a view to a deeper understanding of their dynamic and transient processes. The stages of building numerical models, assessment of their adequacy and specifics in simulations in power electronics are presented. Various methods for modeling current, voltage and resonant inverters in continuous and discontinuous current modes are considered. Specialized models are built in the MATLAB/Simulink and LTspice environments, which factiliate analysis in both steady-state and transient modes. The chapter also contains examples of implementation of series–parallel, parallel–series and combined topologies, applying a unified approach through switching function and differential equations. The goal is to provide a reliable toolkit for engineering prediction, optimization and simulation validation of different types of DC/AC converters.

The fourth chapter develops a unified methodology for the design of DC/AC converters, based on the established models and analytical dependencies. Transfer functions of different types of resonant DC/AC converters operating in hard and soft commutation are presented, including with and without reverse diodes. Complex topologies such as series–parallel, parallel–series and combined resonant circuits are analyzed in detail. Specific methodologies for the design of converters according to the selected structure and operating mode are proposed. The chapter ends with nine real examples for the design of different inverter topologies, including input parameters, simulation results and comparative analysis. This makes it particularly valuable for practical application in engineering tasks and development activities.

The fifth chapter contains the conclusion of the book, which summarizes the main theoretical and applied results of the analysis, modeling and design of DC/AC converters. The importance of applying a unified approach when considering different inverter topologies, which unites physical processes, mathematical models and engineering requirements, is emphasized. The emphasis is on the contribution of the developed methodologies to the creation of highly efficient, reliable and energy-sustainable solutions in the field of power electronics. The chapter also emphasizes future development opportunities, including the application of artificial intelligence and digital twins in the design and optimization of new generation inverter systems.

At the end of the book, a complete list of references is provided for in-depth further study of the topic.

2. Analysis of DC/AC Converters

2.1. Current-Source Inverters: Operating Principle, Variants of the Basic Schematic Topology, and Main Characteristics

2.1.1. Introduction

Current-source inverters (CSIs) have been widely used and commercialized since the 1960s. The main areas of application are industrial technologies and AC electric drives. Their distinctive feature is the large inductance of the input choke included in the DC power supply circuit. Therefore, the current through the semiconductor elements between two consecutive switchings is well smoothed (with small ripples), regardless of the load parameters. Depending on the way the thyristors are turned on, the current inverter circuits are divided into full-bridge and center-tapped transformer configurations. In them, due to the close to rectangular shape of the current in the AC circuit, there is a very good use of the semiconductor elements and therefore the current inverters have a smaller installed power of the equipment compared to resonant inverters. Thanks to the large inductance of the input choke in these circuits, the construction of current overload protection is facilitated. Their disadvantages include the relatively high rate of increase in the current through the semiconductor elements (often, to limit it, a small inductance is included in the circuit of each element). In this aspect, the relatively low maximum operating frequency is also included, which in modern transistors usually does not exceed 100 kHz.

2.1.2. Principle of Operation, Analysis and Characteristics of a Parallel Current-Source Inverter with Active Load

The thyristor version of the single-phase full-bridge CSI is shown in Figure 6. Thyristors VS_1–VS_4 form the inverter bridge, in one diagonal of which the load Z and the output capacitor C are connected in parallel, and in the other diagonal the DC supply voltage U_d is supplied through the choke L_d [5,16,17,19].

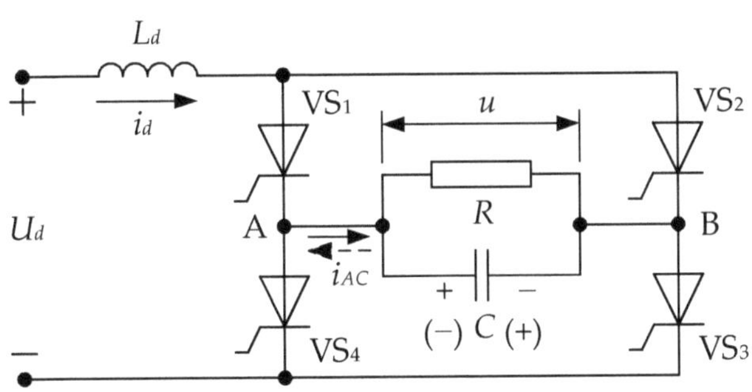

Figure 6. Scheme of a single-phase full-bridge thyristor parallel current-source inverter. Source: Figure by author.

Due to the large inductance of the choke, the input current of the inverter i_d is very well smoothed and has small ripples. Figure 7 shows the timing diagrams illustrating the operation of the CSI in steady state. When control pulses are simultaneously applied to thyristors VS_1 and VS_3 by the inverter control system, current flows through them $i_{VS1} = i_{VS3}$. This current is generated by the input circuit and is divided into two currents in the bridge diagonal: current i_C, which charges the capacitor C with the voltage polarity shown in the figure without brackets, and current i_R, which flows through the load. After the end of this half-period, the other pair of thyristors VS_2 and VS_4 receives control pulses. For a very short time interval (commutation time), when all four thyristors VS_1–VS_4 are turned on, the current in the AC circuit is switched on. After VS_2 and VS_4 are turned on, the capacitor voltage (output voltage u) with a negative sign is applied to thyristors VS_1 and VS_3, which after an angle $\beta = 2\pi f t_{qc}$ (f—control frequency; t_{qc}—thyristor turn-off time), called the turn-off angle (recovery), changes its polarity and becomes positive.

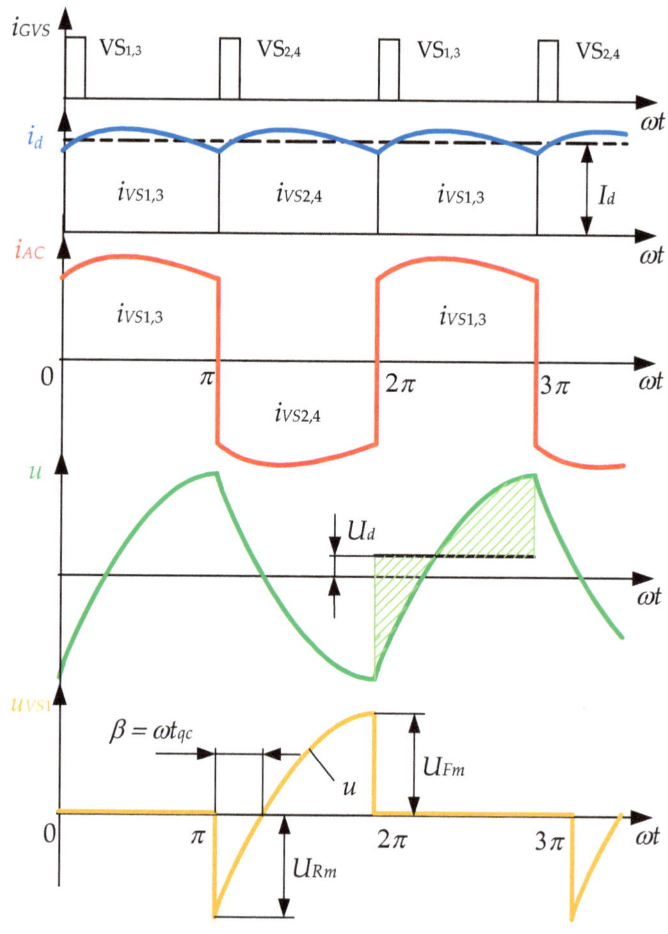

Figure 7. Timing diagrams of the main quantities of a full-bridge CSI at R load. Source: Figure by author.

After the switching moment, the thyristors VS_2 and VS_4 provide a circuit for the current of the DC power source i_d, and the capacitor C is recharged with the reverse polarity shown in brackets in Figure 7. A current flow through the active load in the opposite direction with respect to the current in the previous half-cycle, when VS1 and VS3 were conducting. Then the processes in the circuit are repeated It is possible, in addition to directly, to turn on the load through a transformer in order to match the output voltage u of the inverter and the parameters of the supplied AC load.

The analysis of the full-bridge circuit of the parallel CSI with an active load is performed on the basis of the equivalent circuits shown in Figure 8a,b. The following assumptions were made in the analysis, which do not significantly affect the accuracy of the results obtained: thyristors VS_1–VS_4 are ideal switches with zero resistance in the on state and infinitely large resistance in the off state; the active resistances of the input choke and the power source are zero, and the inverter is powered by a current source without ripples; the inverter has reached a periodic steady state; when the load is connected by a transformer, the transformer is considered ideal, and R and C are the corresponding values of the elements connected to its primary winding.

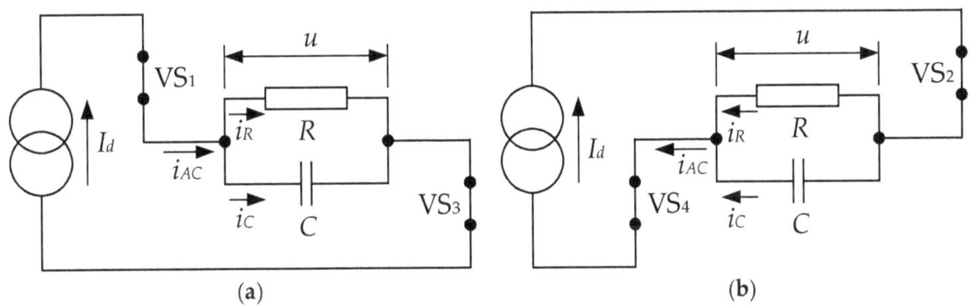

Figure 8. Equivalent circuits of the inverter through the different states of the semiconductor elements: (**a**) with VS_1 and VS_3 turn-on, and VS_2 and VS_4 turn-off; (**b**) with VS_2 and VS_4 turn-on, and VS_1 and VS_3 turn-off. Source: Figure by author.

The equivalent circuits in Figure 8 correspond to the following differential equation:

$$\frac{u}{R} + C\frac{du}{dt} = I_d. \tag{1}$$

The solution to Equation (1) is:

$$\frac{u}{I_d R} = 1 + Ae^{-\frac{t}{\tau}}, \tag{2}$$

where $\tau = RC$ is the time constant of the inverter AC circuit, and A is an integration constant, which is determined by the condition for periodicity of the inverter output voltage u within each half-cycle: $u(-T/4) = -u(T/4)$, where $T = 1/f = 2\pi/\omega$ is the period of the thyristor control pulses. Thus, for the integration constant we obtain

$$A = -\frac{1}{cosh\left(\frac{1}{4fRC}\right)}. \tag{3}$$

After substitution in Equation (1) the solution of the differential equation is:

$$\frac{u}{I_d R} = 1 - \frac{e^{-\frac{t}{RC}}}{cosh\frac{1}{4fRC}}. \tag{4}$$

Due to the properties of the inductance L_d, the average value of the inverter output voltage u for one half-cycle is equal to the voltage of the DC power source U_d (see Figure 7):

$$\frac{2}{T}\int_{-T/4}^{T/4} u \cdot dt = U_d. \quad (5)$$

After performing the conversions, the following relationship between the current and voltage of the DC power source is obtained:

$$U_d = I_d R\left(1 - \frac{4}{B}\, tanh\frac{B}{4}\right), \quad (6)$$

where $B = \frac{1}{fRC}$ is an important circuit parameter called the load factor of the parallel CSI.

Using Equations (4) and (6), the output voltage of the inverter is obtained in relative units, relative to the voltage of the DC power source U_d:

$$u' = \frac{u}{U_d} = \frac{1 - \frac{e^{-fBt}}{cosh\frac{B}{4}}}{1 - \frac{4}{B}\, tanh\frac{B}{4}}. \quad (7)$$

Figure 9 shows the timing diagrams of the inverter output voltage u for two different values of the load factor B. From them it can be seen that with an increase in the value of B (which corresponds to a decrease in the values of C and R), the amplitude value of the voltage u and the recovery interval of the semiconductor elements decrease.

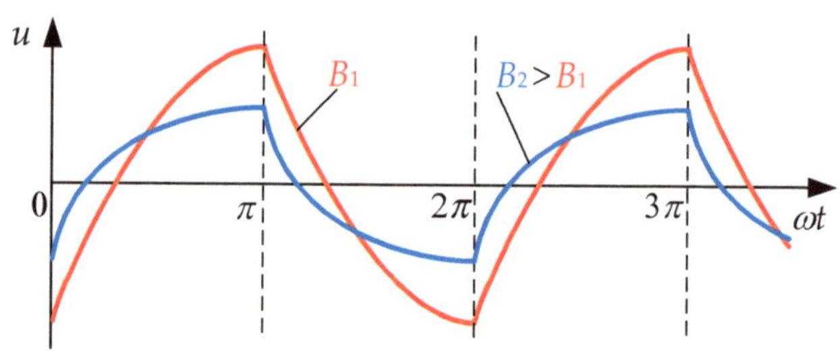

Figure 9. Output voltage of the CSI, at different values of the load factor B. Source: Figure by author.

The RMS value of the inverter output voltage U is determined by the active power in the load P and the load resistance R [5,19]:

$$U = \sqrt{PR}. \quad (8)$$

Assuming that the efficiency of the inverter is $\eta = 1$ (in practice η = 0.94–0.96 $\approx$ 1), the output power is equal to the power consumed by the power source: $P = P_d = U_d I_d$. Then, from the given dependencies, the expression for the external characteristic of the inverter in relative units is obtained [5,20]:

$$U' = \frac{U}{U_d} = \sqrt{\frac{I_d R}{U_d}} = \sqrt{\frac{1}{1 - \frac{4}{B} tanh\left(\frac{B}{4}\right)}}. \quad (9)$$

In a similar way, from (6) the input characteristic of the inverter in relative units is obtained (the dependence of the input current of the inverter on the coefficient B):

$$I_d{}' = \frac{I_d}{U_d f\, C} = \frac{1}{\frac{1}{B} - \frac{4}{B^2} tanh\left(\frac{B}{4}\right)}. \quad (10)$$

The output characteristic of the CSI when operating with an active load is shown in Figure 10a, and the input characteristic is shown in Figure 10b. From the principle of operation of the power circuit, it is clear that the maximum voltage on the semiconductor elements is equal to the maximum value of the output voltage. From Equation (9) the turn-off (recovery) time of the inverter thyristors in relative units is also determined:

$$t_{qc}{}' = \frac{t_{qc}}{T} = \frac{1}{4} - \frac{ln\left(cosh\left(\frac{B}{4}\right)\right)}{B}. \quad (11)$$

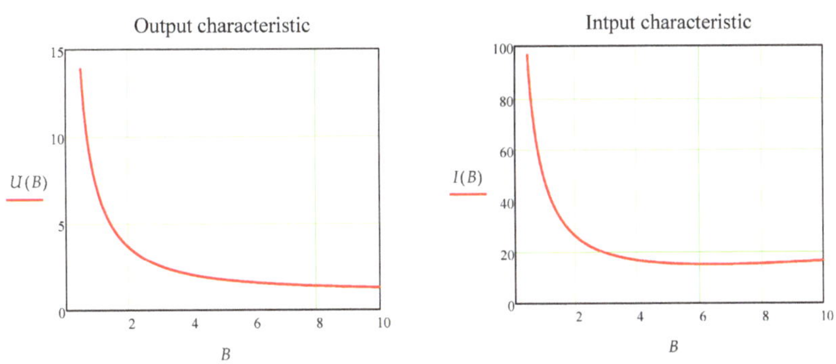

(**a**) (**b**)

Figure 10. Characteristics of a current-source inverter with active load: (**a**) output; (**b**) input. Source: Figure by author.

The scheme of a single-phase parallel current inverter with a midpoint of the output transformer (shown in Figure 11) also has practical applications. In principle, the processes in this inverter do not differ from those in the bridge inverter. When the thyristors VS_1 and VS_2 are switched on again, the output capacitor C is recharged by the supply voltage, periodically changing the polarity of the voltage. The blocking of one thyristor occurs when the other is unblocked, in which case the capacitor C is connected in parallel with the thyristor that needs to be blocked, supplying it with reverse voltage.

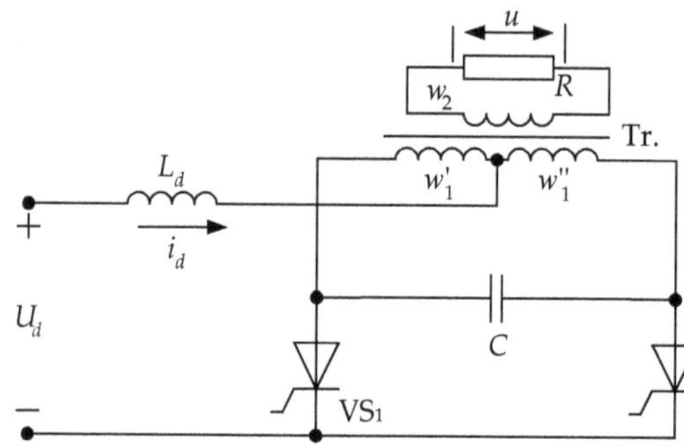

Figure 11. Parallel current-source inverter with transformer midpoint. Source: Figure by author.

2.1.3. Analysis and Characteristics of a Parallel CSI with an Active- Inductive Load

In most cases of parallel CSI applications, the load is active-inductive with a low power factor $cos\varphi_L$. Thus, together with the output capacitor, a parallel load oscillatory circuit is formed with a relatively large quality factor of 3 to 20 (due to the small value of the active component of the load R compared to the inductive ωL) [16,17]. Figure 12 shows the two possible load configurations: a series equivalent circuit (Figure 12a) and a parallel equivalent circuit (Figure 12b).

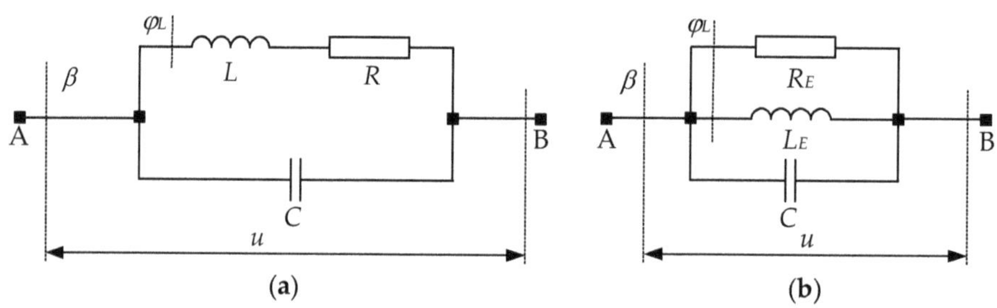

Figure 12. Parallel load circuit: (**a**) with series-connected active load and inductance; (**b**) with parallel-connected active load and inductance. Source: Figure by author.

In this sense, the load circuit acts as a filter for the first harmonic of the inverter output voltage, which is shown in the timing diagrams in Figure 13, which describe the principle of operation of the circuit. In order to ensure sufficient circuit turn-off time of the inverter thyristors and stable switching processes in the inverter, the parallel load circuit has a capacitive detuning, indicated in the figure by the leading angle β.

Due to the filtering properties of the parallel load circuit, the analysis is usually performed using the first harmonic method, with the following assumptions made: the semiconductor elements (thyristors, IGBTs, etc.) are ideal—they have no voltage drop when on, they do not conduct current when off, and the current switching through them is instantaneous; the inductances of the input DC power source and the load are considered linear and constant in time; the active component of the load is modeled by a resistor R, and the inductive component by an inductance L. In addition, the voltage of the DC source is ideal and provides a continuous current.

In this case, the output power is equal to the power consumed by the DC power source [5,16,21]:

$$P_d = U_d I_d = P = UI_{AC(1)} cos\beta = U \frac{4}{\pi\sqrt{2}} I_d cos\beta, \quad (12)$$

where U is the RMS value of the output voltage, $I_{AC(1)}$ is the RMS value of the first harmonic of the current in the inverter's AC circuit, and β is the detuning angle of the parallel load circuit. From the given dependencies, the expression for the inverter's output characteristic in relative units is obtained:

$$U' = \frac{U}{U_d} = \frac{\pi}{2\sqrt{2}} \frac{1}{cos\beta} = TF \frac{1}{cos\beta} = TF\sqrt{1 + tg^2\beta}, \quad (13)$$

where $TF \approx 1.11$ is the transfer function of the CSI.

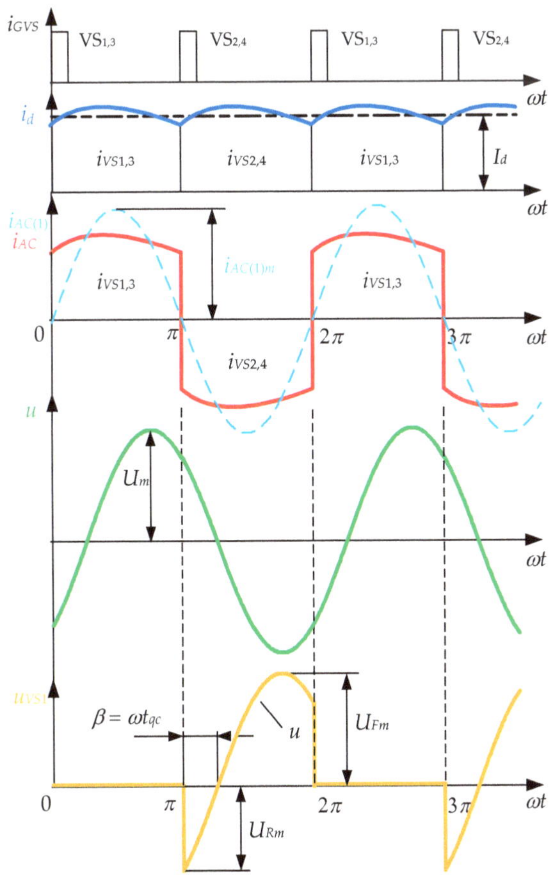

Figure 13. Time diagrams of the main quantities of a full-bridge CSI at *R-L* load. Source: Figure by author.

When designing a current inverter, the load power factor $cos\varphi_L$ is known and therefore it is important to express (13) in terms of the load parameters. For this purpose, the vector diagrams in Figure 14 are used, which describe the relationships between the currents and voltages in the load circuit in a series (Figure 14a) and in a parallel equivalent load circuit (Figure 14b). Using the vector diagrams, the detuning angle of the load circuit is determined as a function of the load power factor:

$$tg\beta = \frac{I_C - I_{LE}}{I_{RE}} = \frac{Ic}{I_{RE}} - \frac{I_{LE}}{I_{RE}} = \frac{U\omega C}{1} \cdot \frac{R_E}{U} - tg\varphi_L = \omega C R_E - tg\varphi_L, \quad (14)$$

where L and R are the active and inductive components of the series equivalent circuit of the load, and L_E and R_E are the active and inductive components of the parallel equivalent circuit of the load. Finally, after conversion, we get:

$$\beta = arctg\left(\frac{1}{B} - tg\varphi_L\right), \quad (15)$$

where $B = \frac{1}{\omega C R_E}$ is the load factor, which has the same form as when operating with a reactive load. After taking into account Equation (15), for the output characteristic of the parallel inverter in relative current units, when operating with *R-L* load, we obtain:

$$U' = \frac{U}{U_d} = TF\sqrt{1 + tg^2\beta} = TF\sqrt{1 + \left(\frac{1}{B} - tg\varphi_L\right)^2}. \quad (16)$$

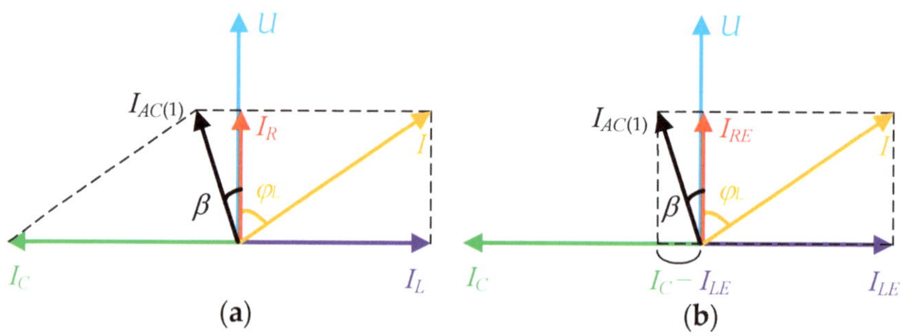

Figure 14. Vector diagrams of the parallel load circuit using different equivalent *R-L* load circuits: (**a**) series; (**b**) parallel. Source: Figure by author.

In Figure 15a, using Equation (16) the output characteristics of a parallel CSI are constructed. From the condition for the balance of active power at the input of the inverter and in the load, we obtain:

$$P_d = U_d I_d = P = UIcos\varphi_L = UI_{RE} = \frac{U^2}{R_E} = U^2 B\omega C. \quad (17)$$

After conversions, the input characteristic of the CSI with an active inductive load is determined:

$$I_d = \frac{U^2}{U_d} B\omega C. \quad (18)$$

Similarly to the output and input characteristics, it is presented in relative units in order to summarize the obtained ratios:

$$I' = \frac{I_d}{\omega C U_d} = TF^2 B\left(1 + tg^2\beta\right) = TF^2 B\left(1 + \left(\frac{1}{B} - tg\varphi_L\right)\right). \tag{19}$$

The input characteristic of the CSI, at different values of the load power factor, is plotted in Figure 15b.

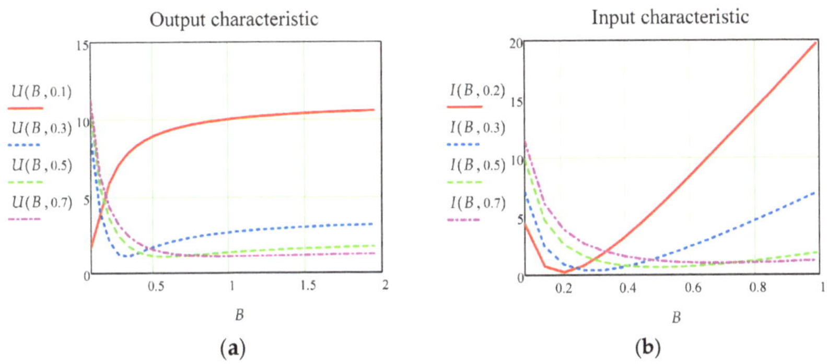

Figure 15. Characteristics of a parallel current-source inverter with *R-L* load: (**a**) output; (**b**) input. Source: Figure by author.

From the point of view of the design of the current-source inverter, the relationship between the parameters of the series and parallel equivalent circuits of the *R-L* load is of interest. This relationship is conveniently found using the resistance triangle given in Figure 16a and the conductivity triangle (Figure 16b). Usually, the output active power P, the RMS value of the output voltage U, the operating frequency f and the power factor of the load $cos\varphi_L$ are known. From these quantities, the value of the active resistance of the parallel equivalent circuit of the load is directly determined:

$$R_E = \frac{U^2}{P}. \tag{20}$$

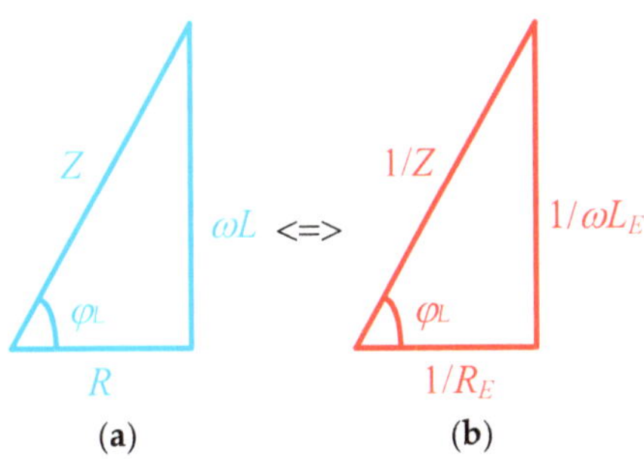

Figure 16. Equivalent circuits for determining the parameters of the equivalent load circuit of a parallel current-source inverter with *R-L* load: (**a**) with impedance triangle; (**b**) with conductance triangle. Source: Figure by author.

According to the equivalent circuits in Figure 16, the power factor of the load is determined by:

$$cos\varphi_L = \frac{R}{Z} = \frac{1/R_E}{1/Z} = \frac{Z}{R_E} \text{ or } RR_E = Z^2. \tag{21}$$

After transformations, the result is that

$$R_E = R\left(1 + tg^2\varphi_L\right) \text{ or } R = R_E cos^2\varphi_L. \tag{22}$$

From the triangles shown in Figure 16, the following is obtained:

$$sin\varphi_L = \frac{X_L}{Z} = \frac{1/X_{LE}}{1/Z} = \frac{Z}{X_{LE}} \text{ or } X_L X_{LE} = Z^2. \tag{23}$$

Similarly, the relationship between the reactive components of the load from the different equivalent circuits is determined:

$$X_{LE} = X_L\left(1 + cotg^2\varphi_L\right) \text{ or } X_L = X_{LE} sin^2\varphi_L. \tag{24}$$

The quantities required for designing a CSI with an *R-L* load are determined by similar dependencies as for an active load.

2.2. *Voltage Source Inverters*

2.2.1. Introduction

Voltage source inverters (VSIs) are an important class of power electronic devices that are widely used in various fields of electronics and energy—from domestic and industrial power supplies to photovoltaic systems, electric vehicles and uninterruptible power supplies (UPSs) [5,13,20,21]. The main role of the VSI is to provide an output voltage with a certain amplitude and frequency, which is compatible with the electrical loads. The direct current from the source (e.g., battery, rectified voltage from the power supply network or DC/DC converter) is interrupted and directed through various electronic switches (most often pairs of transistors), which are controlled with appropriate logic. In this way, alternating half-cycles of positive and negative voltage are created at the output of the device.

VSIs are classified according to various features [1,2,4,22]:

- According to the circuit design: Half-bridge inverters use two main switches and two separation capacitors. They have a simpler configuration, but require additional elements for voltage balancing. In addition, separation capacitors have a large capacitance value and are in many cases more expensive and unreliable compared to the use of semiconductor elements. Full-bridge inverters use four semiconductor switches, through which a complete change in the output polarity is achieved. They are more efficient and widely used. Multilevel inverters implement more than two levels of the output voltage by using more semiconductor switches, which achieves a better quality of the output voltage with lower harmonic distortions and switching losses.
- According to the number of phases of the output voltage: single-phase; three-phase and multiphase.

- According to the shape of the output voltage: Inverters with square wave voltage—simple and cheap, but with a high content of harmonics. Inverters with modified sine wave—an improved version that imitates a sine wave through a step signal (an improved version of those with a rectangular shape, characterized by less harmonic distortion). Inverters with pure sine wave—the output signal is practically identical to that of the power distribution network; they are used to power sensitive consumers. Pulse-width modulation (PWM): PWM is the most common method for creating a sinusoidal output voltage in VSI. In this method, electronic switches are switched at a high frequency, adjusting the width of the pulses so that the average voltage value follows a sinusoidal shape. By filtering the PWM signal (using LC filters), the high-frequency components are removed and a pure sine wave is obtained at the output.

The advantages of the PWM method are as follows [8,9,23]: better voltage quality; ability to adjust amplitude and frequency; higher efficiency and low losses. In the implementation of the VSI, both fully controlled semiconductor devices (transistors, dual-operation thyristors) and single-operation thyristors with the corresponding switching circuits are used.

VSIs are a key part of modern energy and electronic systems. By choosing the right circuit solution and control (especially with the application of PWM technology), a very good shape of the output voltage and high efficiency are achieved. The constant development of power electronics makes these devices increasingly reliable, compact and accessible for mass application [24–26].

2.2.2. Analysis of a Single-Phase Full-Bridge Voltage Source Inverter

Single-phase VSIs are most often implemented in a full-bridge circuit, as shown in Figure 17. Usually the load is active-inductive and is included in the output diagonal of the bridge, consisting of transistors VT_1–VT_4. Diodes VD_1–VD_4 are connected in parallel to each transistor and create a path for the load current to pass when it is in the opposite direction to the current of the transistors. The DC power source with a value of U_d is included in the other diagonal of the bridge. When considering the principle of operation and analyzing the circuits, it is assumed that the devices are ideal switches with unidirectional conductivity and the DC power source has zero internal resistance.

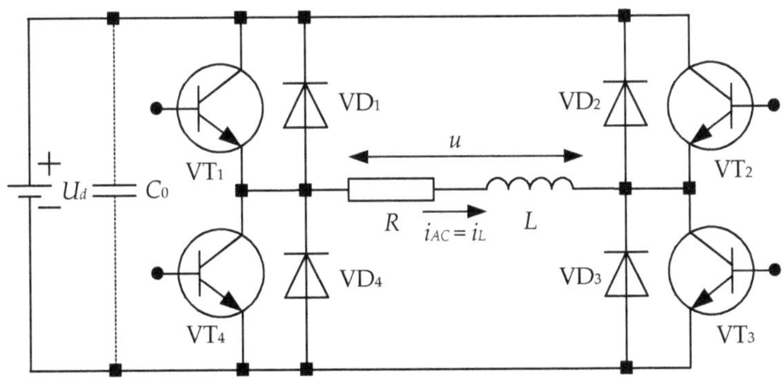

Figure 17. Full-bridge single-phase voltage source inverter. Source: Figure by author.

The principle of operation of the circuit is explained by the timing diagrams shown in Figure 18. At time t_1 let control pulses be applied to transistors VT_1 and VT_3. From the power source through the uncapped transistors and the load, a current with an exponential form flows, and a voltage with positive polarity and amplitude U_d is applied to the load. At time t_2, transistors VT_1 and VT_3 are blocked, and control pulses are applied to the transistors on the other diagonal of the bridge VT_2 and VT_4. Due to the active-inductive nature of the load, the current i retains its direction for some time until time t_3, depending on the time constant of the load $\tau = L/R$. In the interval t_2–t_3, the flow of the load current is ensured by the energy stored in the inductance L, which returns to the power source through the reverse diodes VD_2 and VD_4. As a result of the operation of the diodes, the voltage on the load changes its polarity, but remains with the same value—U_d. At time t_3, the current i drops to zero and since control pulses are still applied to transistors VT_2 and VT_4, they turn on, and the load starts consuming energy from the power source again. The control and blocking of VT_2 and VT_4 stop at time t_4, where control pulses are applied to the other pair of transistors VT_1 and VT_3. VT_1 and VT_3 cannot turn on until the load current changes direction, which happens at time t_5. In the interval t_4–t_5, diodes VD_1 and VD_3 ensure the passage of the reactive current of the load after the switching of the transistors, and as a result of their operation, the load voltage changes its polarity again. After time t_5, VT_1 and VT_3 turn on and the processes in the circuit are repeated. Capacitor C_0 provides the zero internal resistance of the DC power source in order to ensure the return of energy from the load.

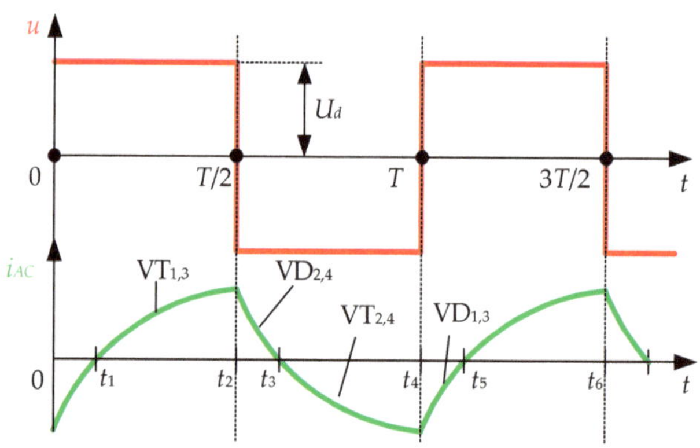

Figure 18. Timing diagrams describing the operating principle of a full-bridge single-phase voltage source inverter. Source: Figure by author.

The load voltage u is rectangular in shape and is decomposed into a Fourier series containing n harmonic components:

$$u(t) = \frac{4U_d}{\pi}\left(sin(t) + \frac{1}{3}sin(3t) + \frac{1}{5}sin(5t) + \cdots + \frac{1}{n}sin(nt)\right). \quad (25)$$

as the first (fundamental) harmonic of the output voltage $u_{(1)}(t) = \frac{4U_d}{\pi}sin(t)$ has an amplitude of $U_{(1)m} = \frac{4U_d}{\pi} = 1.27U_d$ and an RMS value of $U_{(1)} = \frac{4U_d}{\pi\sqrt{2}} = 0.9U_d$.

From Equation (25) it is established that the amplitude of the 3rd harmonic is 33.3%, the 5th harmonic is 20%, and the 7th harmonic is 14.3%, of the amplitude of the fundamental. If it is necessary to obtain a load voltage with a sinusoidal shape, methods must be used to eliminate the remaining harmonic components of the voltage, such as using a filter between the load and the inverter and/or applying PWM to form the control pulses to the semiconductor switches. In the considered method of forming the inverter voltage, in which the transistors are controlled with pulses with a duration equal to the half-period of the output voltage, regulation of the voltage value is possible only by changing the magnitude of the voltage of the DC power source U_d.

The electromagnetic processes in the power diagram of Figure 17 correspond to the following differential equation:

$$L\frac{di_L}{dt} + Ri_L = U_d. \tag{26}$$

From the comparison of Equations (1) and (26), a duality is established in the description of the processes in single-phase CSI and VSI as follows: the current source I_d, the voltage of the capacitor and the load u and parallel capacitor C from (1) are replaced, respectively, by a voltage source with a value U_d, current through the inductance and the load i_L and a series inductance L from Equation (26).

The solution of Equation (26) is the following [1,5]:

$$i_L(t) = \frac{U_d}{R} + A\,e^{-\frac{t}{\tau}}, \tag{27}$$

where $\tau = L/R$ is the time constant of the inverter AC circuit, and A is an integration constant, which is determined by the condition for periodicity of the current through the load i_L within each half-cycle: $i_L(-T/4) = -i_L(T/4)$, and $T = 1/f$ is the period of the transistors control pulses. Thus, for A we obtain:

$$A = -\frac{2U_d}{R}\frac{1}{1+e^{-\frac{1}{2f\tau}}}. \tag{28}$$

After substituting the thus determined constant into (27) the solution of the differential equation is:

$$i_L(t) = \frac{U_d}{R}\left(1 - \frac{2e^{-\frac{t}{\tau}}}{1+e^{-\frac{1}{2f\tau}}}\right). \tag{29}$$

The maximum value of the current in the AC circuit occurs at the end of the half-cycle and is determined by:

$$I_{Lm}(t) = \frac{U_d}{R}\left(1 - \frac{1-e^{-\frac{1}{2f\tau}}}{1+e^{-\frac{1}{2f\tau}}}\right). \tag{30}$$

With the assumptions made in the analysis, the maximum value of the voltages on the semiconductor devices is equal to the value of the voltage of the input DC power source U_d. The RMS value of the voltage on the load is of the same magnitude.

2.2.3. Methods for Forming and Regulating the Output Voltage in Voltage Source Inverters

The most commonly used method for forming and regulating the output voltage is the pulse-width modulation method (PWM). When using the PWM method for forming and regulating the voltage in the VSI, the output voltage within its period is composed of N pulses of equal or different duration, which can change according to a certain law (for example, sinusoidal). The output voltage is regulated by changing the duration of the pulses.

PWM with Output Voltage Shape Depending on Load Parameters

In this type of PWM, the duration of the transistors' conduction is changed. The principle of operation will be explained based on the bridge circuit of the VSI from Figure 17. In the time interval from t_1–t_2, as shown in the timing diagrams from Figure 19, unclogging pulses are applied to the transistors VT_1 and VT_3, as energy is consumed from the power source and a rectangular voltage with a value U_d is formed on the load. Unlike the considered unregulated VSI, when at the moment t_2 the operating VT_1 and VT_3 are clogged, in the considered algorithm of operation, control of the next transistor pair VT_2 and VT_4 is not applied.

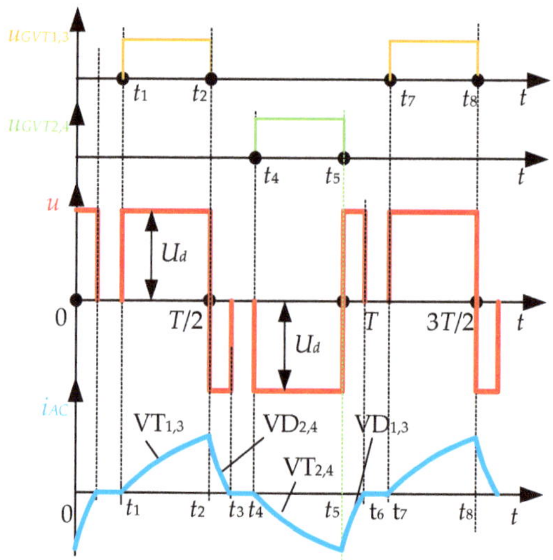

Figure 19. Timing diagrams describing the operating principle of a full-bridge adjustable single-phase voltage source inverter using PWM with an output voltage shape dependent on the load parameters. Source: Figure by author.

The energy stored in the inductance L ensures the flow of current through the reverse diodes VD_2 and VD_4, as at the moment t_3 the load current drops to zero, and in the interval t_2–t_3 the load voltage changes its sign (inverts), but retains its amplitude. Transistors VT_2 and VT_4 are unblocked at time t_4 and thus intervals are obtained in the operation of the DC/AC converter, in this case from t_3–t_4, in which none of the transistors and diodes work, no current flows through the load and there is no voltage on the load.

When the transistors VT_2 and VT_4 are turned on (t_4–t_5), current begins to be consumed from the power source again, and a rectangular voltage with negative polarity is formed on the load. After the blocking of VT_2 and VT_4, through the diodes VD_1 and VD_3, energy begins to return to the power source, and the voltage at the inverter output changes its polarity again. At the moment t_6 the load current and the output voltage become equal to zero and this state continues until the moment t_7, when the transistors VT_1 and VT_3 are unblocked and thus the processes in the circuit are repeated. The conduction of the reverse diodes (intervals t_2–t_3, t_5–t_6, etc.) leads to the appearance of additional pulses in the inverter output voltage curve and hence to an undesirable increase in its RMS value. The duration of the additional pulses depends on the time constant of the load circuit $\tau = L/R$, which when changing the load leads to a strong dependence of the output voltage value on the load parameters. In practice, cases are possible when the pause time is not enough to dissipate the energy stored in the inductance and the inverter starts to operate as unregulated, without pauses in the output voltage.

PWM with Output Voltage Shape Independent of Load Parameters

A shape of the output voltage independent of the load parameters is achieved if, in the intervals when the output voltage should be zero, simultaneous conduction of two transistors of the inverter bridge is ensured, where the transistors have common electrodes—for example, VT_1 and VT_2 or VT_3 and VT_4 for the circuit in Figure 17—and the load turns out to be short-circuited across the "+" or "−" bus of the power source. The timing diagrams illustrating this method of VSI control are shown in Figure 20.

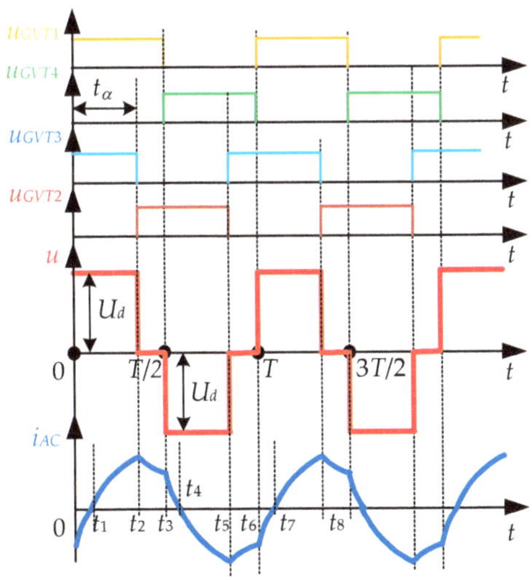

Figure 20. Timing diagrams describing the operating principle of a full-bridge adjustable single-phase voltage source inverter using PWM with an output voltage shape independent of the load parameters. Source: Figure by author.

Each transistor is switched on for an angle of 180°, corresponding to the half-period of the output voltage. The transistors from each of the arms of the bridge, where the load VT_1 and VT_4, and VT_2 and VT_3 is switched on in the same sequence as in the unregulated inverter, with the unblocked state of one corresponding to the blocked state of the other and vice versa. The difference is that the control creates a phase shift t_α during the sequential switching of the transistors from the two diagonals of the bridge. The load current is formed in the following way: in the interval t_1–t_2, VT_1 and VT_3 operate, followed by t_2–t_3, VT_1 and VD_2; t_3–t_4, VD_2 and VD_4; t_4–t_5, VT_2 and VT_4; t_5–t_6, VT_4 and VD_3; t_6–t_7, VD_1 and VD_3; and t_7–t_8, VT_1 and VT_3. The considered control algorithm allows us, by changing the phase shift time t_α from 0 to $T/2$, to regulate the RMS value of the DC/AC converter output voltage from its maximum value to zero. The amplitude of the harmonic components of the output voltage is

$$U_{m(n)} = \frac{4U_d}{n\pi} sinn \frac{t_\alpha}{2}, \tag{31}$$

where n is the number of the next harmonic. To improve the harmonic composition of the output voltage, it is advisable to use PWM with a larger number of pulses within the half-period.

Formation of DC/AC Converter Voltage with Selective Harmonic Content

The method for selective elimination of harmonic components from the load voltage avoids the presence of many components in the harmonic composition with a rectangular shape of the output voltage and the large switching losses in the sinusoidal PWM (SPWM). When using this method, the desired harmonic composition of the output voltage is achieved with a smaller number of switchings of the power electronic elements. The number of switchings and the angles at which they are performed are selected so as to eliminate certain harmonics.

The method under consideration is implemented through the use of bipolar PWM, in which the pauses in the output voltage curve are replaced by pulses of opposite polarity. Such a type of PWM is shown in Figure 21.

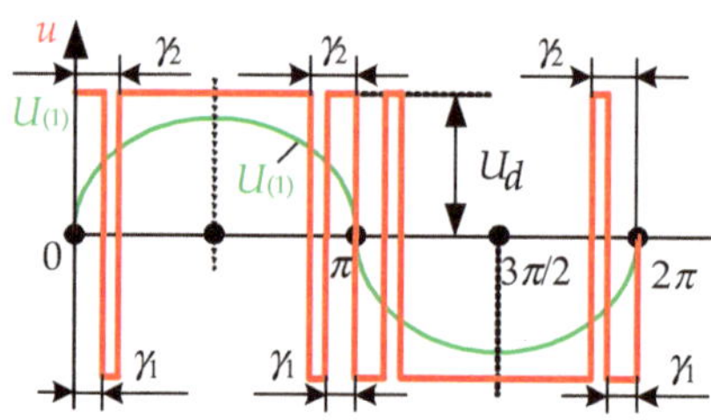

Figure 21. Output voltage formation at VSI with selective elimination of harmonics. Source: Figure by author.

And in this case, the regulation of the RMS value of the output voltage is performed by changing the duration of the pulses. For example, at values of angles γ_1 = 23.62° and γ_2 = 33.3° in the output voltage curve there are no 3rd and 5th harmonics, and at γ_1 = 16.25° and γ_2 = 22.07°—5th and 7th [5]. With this method of operation, the use of the supply voltage is about 5% worse than with the rectangular shape and about 19% better than with the SPWM. The disadvantage is that with each change in the output voltage, in order to preserve the harmonic composition, different ratios between the angles γ_1 and γ_2 are required.

Single-Phase Voltage Source Inverters with Sinusoidal Pulse-Width Modulation of the Output Voltage

The content of harmonic components in the output voltage of the inverter is significantly reduced by using SPWM, in which the DC/AC converter output is formed from a series of rectangular pulses, the duration of which varies according to a sinusoidal law. With SPWM, a predominant content of the first harmonic in the output voltage curve is obtained and a minimal presence of harmonics close to the fundamental (3rd, 5th, 7th, etc.) [1,13]. The adjustment of the RMS value of the output voltage is performed by changing the duration of the pulses, which is changed by the depth of modulation. When using different topologies, there is specificity in the implementation of SPWM in the full-bridge circuit (Figure 17), the half-bridge circuit (Figure 22a) and the circuit with a midpoint of the output transformer (push–pull circuit) (Figure 22b).

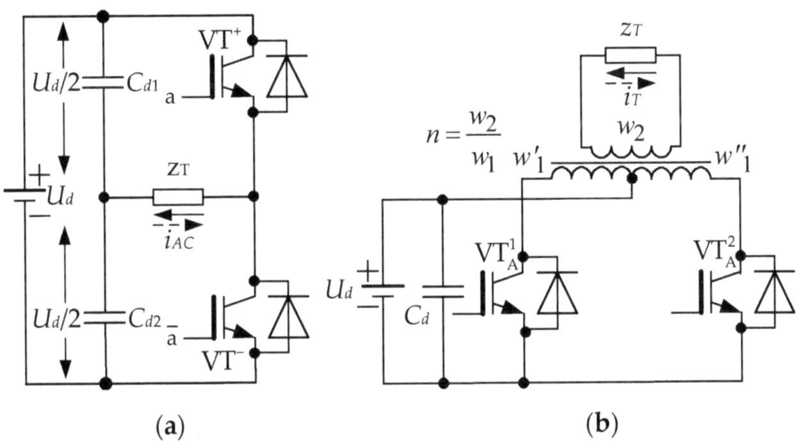

Figure 22. Schematic variations of voltage source inverters: (**a**) half-bridge; (**b**) with midpoint of the output transformer. Source: Figure by author.

In the half-bridge circuit (Figure 22a), the demonstrated modulation is bipolar, since during one half-cycle the output voltage of the DC/AC converter changes its polarity. The switching on and off of the power semiconductor elements is performed by comparing in the control system the control voltage u_{Contr}, which is with the frequency of the first harmonic component of the output voltage (corresponding to a period $T_{(1)}$ and the sawtooth voltage u_{Tr} (with a period T_S), as shown in Figure 23). If the control voltage is greater than the sawtooth, the upper power electronic element, marked in Figure 22a with VT^+, is switched on, and if it is less, the lower VT^-. When the corresponding key element is switched on, the voltage on the load is equal to half the voltage value of the DC power source. The value of the first harmonic of the output vFoltage is adjusted by changing the ratio of the amplitudes of the control and sawtooth signals $M_A = U_{Contrm}/U_{Trm}$. The value of the coefficient M_A can be changed between 0 and 1, and for this purpose the amplitude of the control voltage U_{Contrm} is changed. The first harmonic of the output voltage $U_{(1)}$ and its maximum value $U_{(1)m}$ are marked in Figure 23.

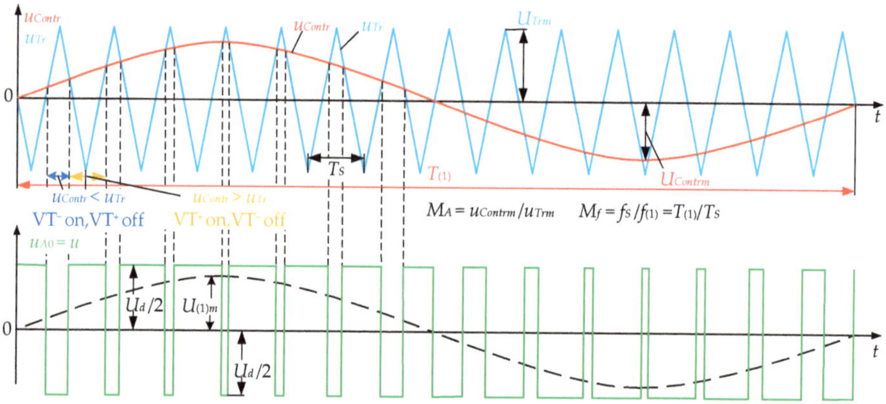

Figure 23. Bipolar SPWM for half-bridge circuit of the VSI. Source: Figure by author.

Since the switching of the transistors is at a high frequency (usually over 100 times compared to the frequency of the output voltage), it is assumed that for the relatively small period T_S the value of the control voltage u_{Contr} does not change [20]. For the same reason, it is assumed that the instantaneous value of the first harmonic of the output voltage is equal to the average value for this interval, obtained, respectively, as the sum of one positive and one negative voltage, as shown in Figure 23.

Since the control voltage varies according to a sinusoidal law with the frequency of the first harmonic of the output voltage, the average value of the inverter output voltage as a function of time is obtained as follows:

$$\overline{u}(t) = \frac{U_{Contrm} \cdot sin\omega_{(1)}t}{U_{Trim}} \cdot \frac{U_d}{2} = M_A \cdot \frac{U_d}{2} \cdot sin\omega_{(1)}t. \quad (32)$$

Therefore, the maximum value of the first harmonic of the output voltage depends on the value of the supply and on the ratio of the amplitudes.

$$U_{(1)m} = \frac{U_{Contrm}}{U_{Trim}} \cdot \frac{U_d}{2} = M_A \cdot \frac{U_d}{2}. \quad (33)$$

An important parameter determining the harmonic composition of the DC/AC converter output voltage is the ratio of the frequencies of the reference sine wave and the sawtooth voltage $M_f = f_S/f_{(1)}$, which in the implementation of bipolar modulation must be an odd number. For the ratio of the amplitudes in Figure 24, the distribution of the amplitudes of the individual harmonics in relative units, relative to the input voltage of the inverter, is given. From the graph it is established that the harmonic components with a small number (3rd, 5th, 7th), characteristic of a rectangular shape of the output voltage, are absent [2,27]. When the switching frequency of the power semiconductor elements increases, i.e., when the value of the coefficient M_f increases, the harmonics are located at higher frequencies and their filtering is facilitated.

When adjusting the output voltage (by changing the value of M_A), the values of the harmonic components change. The largest is the value of the harmonic, whose frequency is M_f times greater than the fundamental. At M_A = 0.8 it is almost equal to the fundamental, but at small values of M_A it may turn out to be with a greater value than the fundamental harmonic.

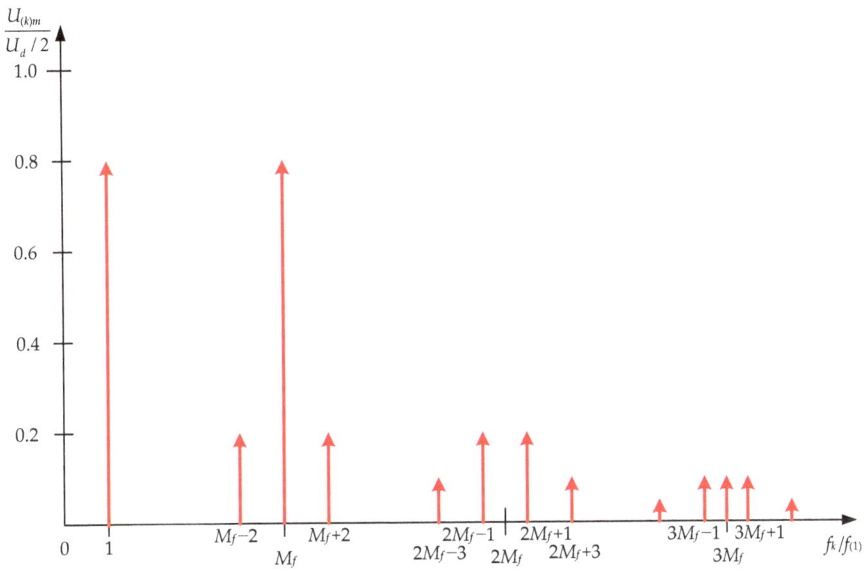

Figure 24. Harmonic composition of the output voltage of the VSI with bipolar PWM, with a value of the coefficient M_A = 0.8. Source: Figure by author.

The bipolar SPWM can also be implemented in the bridge circuit of the VSI (Figure 17). The operation of the DC/AC converter is as follows: when the control voltage is with a greater value than the sawtooth, VT_1 and VT_3 are turned on, otherwise VT_2 and VT_4. When a pair of power electronic elements is turned on, the entire supply voltage is applied to the load. The first harmonic is with the frequency of the output voltage and its maximum value is determined by Equation (33). Due to the peculiarities of the power circuit, it is twice as large as that in the half-bridge circuit. The SPWM improves the harmonic composition of the output voltage, but worsens the use of the supply voltage compared to the case with a rectangular shape.

Further improvement of the harmonic content can be achieved by using a unipolar SSIM, shown in Figure 25, when implementing a full-bridge circuit of the VSI.

In this type of modulation, the coefficient M_f must be an even number. The principle of operation is as follows: when the control voltage is greater than the sawtooth voltage, VT_1 is turned on, otherwise VT_4. When the control voltage shifted by 180° is greater than the sawtooth voltage, VT_2 is turned on, otherwise VT_3 [2,21]. There are intervals in which either two upper or two lower power semiconductor elements are simultaneously turned on. These are the intervals of pauses in the output voltage, during which one power key element conducts with the reverse diode of its neighbor, connected to the same pole of the supply voltage. The dependence (33) for the maximum value of the first harmonic of the output (load) voltage is also valid in this case, taking into account the fact that the entire voltage of the input DC power source U_d is converted. Figure 26 shows the harmonic composition in unipolar SPWM. The harmonic components are shifted towards higher frequencies, which makes their filtering easier. Compared to Figure 24, the harmonics around M_f are absent, then again around $3M_f$, etc.

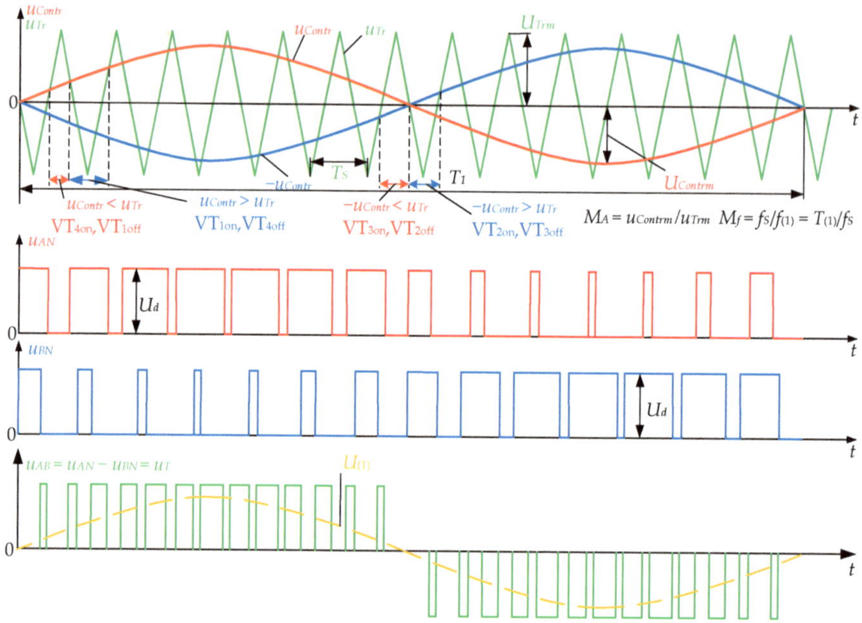

Figure 25. Unipolar SPWM for VSI full-bridge circuit. Source: Figure by author.

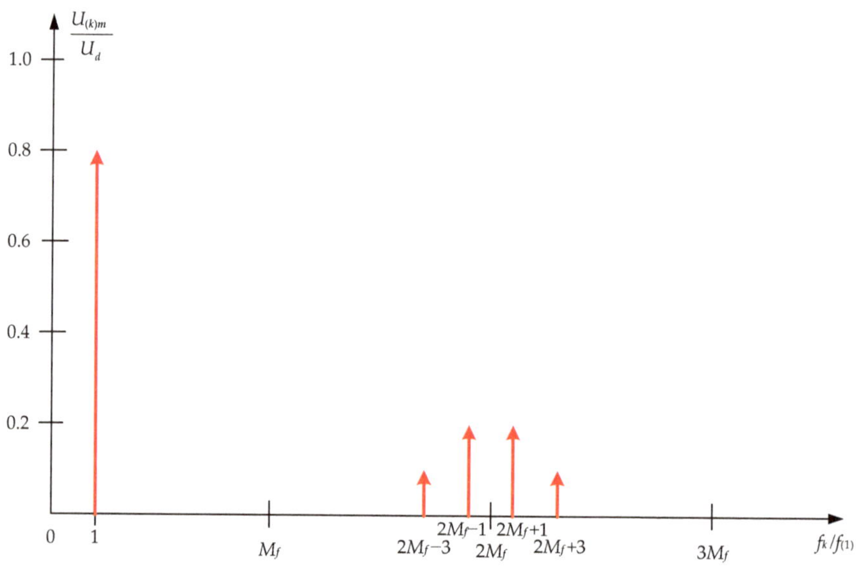

Figure 26. Harmonic composition of the output voltage of the VSI in unipolar SPWM, with a value of the coefficient $M_A = 0.8$. Source: Figure by author.

2.2.4. Three-Phase Voltage-Source Inverters

Three-phase VSI can be implemented by three single-phase VSI operating with a phase difference of 120° in them or by using a three-phase full-bridge circuit, as shown in Figure 27, and the load can be connected in a "star" or " triangle" configuration.

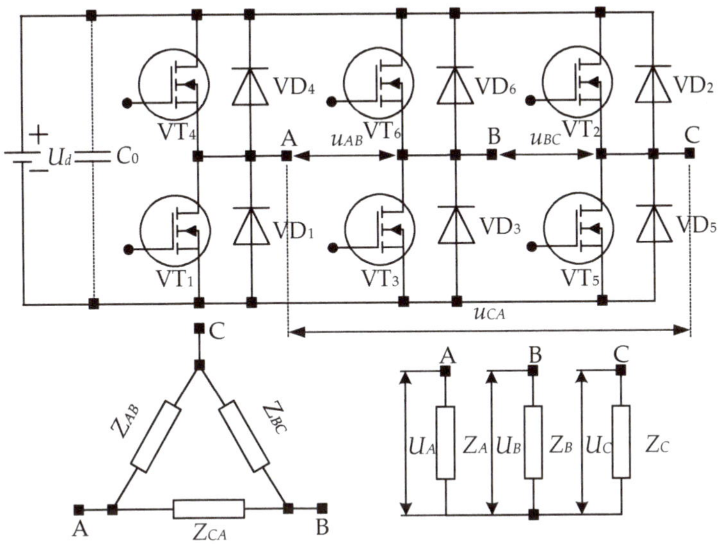

Figure 27. Three-phase transistor voltage source inverter with different load connection configurations. Source: Figure by author.

One of the possible algorithms of operation of a three-phase VSI is illustrated in Figure 28. At any given moment, three transistors are simultaneously conducting, two of which are from the same group. For example, the phase-to-phase voltage u_{AB} is positive and equal to the supply U_d when transistors VT_3 and VT_4 are conducting together, it is negative when VT_1 and VT_6 are operating, and it is equal to zero when VT_1 and VT_3 or VT_4 and VT_6 are open.

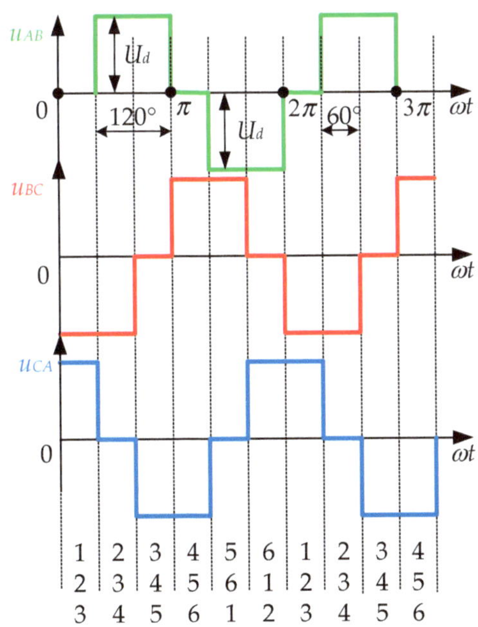

Figure 28. Example of three-phase voltage formation. Source: Figure by author.

When using single-operation thyristors for the implementation of the VSI, it is necessary to add additional switching elements to provide reverse voltage to turn off the operating thyristors. With the development of the element base of power electronics, the use of single-operation thyristors in VSI is becoming increasingly rare at the expense of powerful IGBT and MOS transistors.

2.3. Resonant DC/AC Converters

2.3.1. Introduction

The development of modern power electronics is a continuous effort to increase efficiency, reduce losses and improve the reliability of power electronic devices. In this context, resonant inverters (RIs) represent one of the most innovative and effective solutions for converting electrical energy using electronics, in applications requiring high frequency and low switching losses [1,22,28,29]. Unlike traditional "hard-switching" inverters, in which the switching of key elements leads to significant power losses, RIs use the principle of "soft-switching". This allows the switching of power semiconductor elements to be performed at zero voltage (zero-voltage switching, ZVS) and/or at zero current (zero-current switching, ZCS), which significantly reduces thermal load and extends the life of the components [18,30,31].

The basic principle of operation of RIs is the use of LC resonant circuits of varying complexity, which allow the formation of currents and voltages with a shape close to sinusoidal [17]. By properly choosing the resonant frequency and the parameters of the circuit, optimal operation of the inverter is achieved with minimal losses. Thanks to this feature, RIs are used in a wide range of applications—from induction heating and pulsed power supplies to wireless energy transmission and high-frequency medical devices [20]. Another important advantage is the possibility of RI operation at high frequencies, which leads to a reduction in the size of passive elements and facilitates the construction of compact and lightweight devices. In addition, due to the specific nature of their operation, they provide a lower level of electromagnetic interference (EMI), which facilitates the electromagnetic compatibility of devices [2,21,32].

RIs are classified according to the configuration of their resonant circuit and the topology of the power circuit. The main types are [1,5,6,19]:

- Series Resonant Inverter (SRI). In this type of inverter, the resonant circuit consists of inductance and capacitance connected in series with the load. When the resonant frequency is reached, the current in the circuit becomes maximum, allowing efficient energy transfer. This type of inverter is particularly suitable for applications with variable loads, such as induction heating.
- Parallel Resonant Inverter (PRI). In this variant, the resonant circuit is connected in parallel with the load. This allows a stable voltage across the load under alternating current. Parallel resonant inverters are often used in cases where a stable output power is required regardless of changes in the load.
- Series–Parallel Resonant Inverter (SPRI). This type combines the advantages of the previous two, using a combined resonant circuit. Such schemes allow better adaptation to different types of loads and offer flexibility in adjusting the output parameters;
- RI with a complicated output circuit. This is a power electronic converter in which the process of inverting electrical energy is implemented by using a

multi-component resonant structure in the output circuit. This structure usually includes a combination of inductive and capacitive elements configured in series, parallel or combined (series–parallel) resonant loops that interact with the load and provide conditions for soft switching. Examples of such RIs are LLC, LCC and LCLC.

- Depending on the configuration of the power section, inverters are full-bridge, split-resonant or half-bridge RI. Full-bridge circuits allow for greater output power, while half-bridge and split-resonant capacitor inverters offer a simpler structure and reduced costs.
- Multi-unit RI. This is a type of RI in which the device is built from two or more series or parallel connected resonant units (modules), operating in concert to improve energy efficiency, expand the frequency range and optimize operation under variable loads. Each unit can be designed with different resonant frequencies, inductances and capacitances, so that the overall system meets specific requirements for output voltage, current and dynamic characteristics. Multi-unit RIs allow for more precise power control, as well as the ability to implement self-adaptation to load requirements and control harmonic components.

RIs most often contain a series RLC circuit at their output, consisting of resonant inductance L, resonant capacitor C and load R. When the condition for resonance of the RLC circuit is met, the current in the AC circuit and the output voltage have a shape close to sinusoidal. Operation in an oscillatory (resonant) mode is ensured when the following condition is met [1,5]:

$$R < 2\sqrt{\frac{L}{C}}. \tag{34}$$

RIs are the highest frequency inverters because a current with a shape close to the sinusoidal one flows through the semiconductor devices, which is characterized by a slow increase in current and, accordingly, operation with small switching losses. On the other hand, changes in the load parameters have a strong influence on the operation mode of RIs and this requires the use of different circuit solutions (for example, the inclusion of reverse diodes) and automatic control systems in order to improve their performance [5,20].

The main parameters of the resonant circuit are the natural (resonant) frequency: $\omega_0 = \sqrt{\frac{1}{LC} - \delta^2}$ and the damping: $\delta = \frac{R}{2L}$.

When performing the analysis of resonant DC/AC converters, the following parameters were applied:

- Coefficient of oscillation: $k = \frac{1}{1-e^{-\frac{\delta\pi}{\omega_0}}}$;
- Detuning factor of the equivalent AC circuit ν, which is determined by the ratio of the circular control frequency $\omega = 2\pi f$, to the resonant frequency ω_0: $\nu = \frac{\omega}{\omega_0}$.

In accordance with the long-standing practice of the Department of Power Electronics at the Technical University of Sofia, Bulgaria, the coefficients k and ν are employed as fundamental parameters in the analysis of resonant DC/AC converters [15,20]. Under the adopted definitions, the functional relationship between the oscillation coefficient k and the quality factor Q of a series resonant circuit is rigorously established, enabling its precise and unambiguous determination:

$$Q = \frac{\omega_0 L}{R} = \frac{\omega_0}{2\delta} = \frac{\pi}{2ln\left(\frac{k}{k-1}\right)} \tag{35}$$

2.3.2. Principle of Operation and Basic Ratios

The principle of operation and analysis of the RI will be made on the basis of a half-bridge thyristor circuit shown in Figure 29a. The steady-state operation is considered.

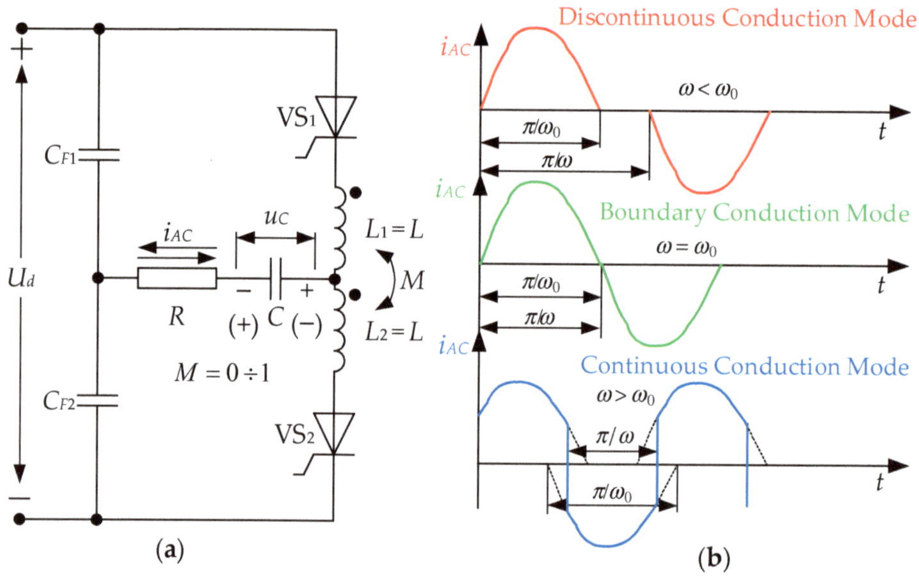

Figure 29. Visualization of the basic circuit and the operating modes of the RI: (**a**) half-bridge thyristor RI; (**b**) operating modes of the RI depending on the detuning of the AC circuit. Source: Figure by author.

Figure 29b shows the three possible operating modes of the RI. The discontinuous conduction mode (DCM) and the boundary conduction mode (BCM), where $\nu \leq 1$ are combined under the general name of the natural commutation mode (NCM) of the devices, because the semiconductor devices are switched off by the decrease to zero in the inverter current [33–35]. In the continuous conduction mode (CCM, $\nu > 1$), also known as the hard commutation mode (HCM), the inverter current is transferred between the devices before it has dropped to zero and for normal operation in this mode it is necessary to take measures to ensure a current path during the switching of the devices. When a control pulse is applied to the thyristor VS_1, it is unblocked and a current i flows through the "+" circuit of the power source, the inductance L_1, the resonant capacitor C, the load R and the common point of the filter capacitors C_{F1} and C_{F2}, which, when the condition for the resonant mode of operation of the RLC circuit is met, has a shape close to the sinusoidal one, as shown in the timing diagrams in Figure 30.

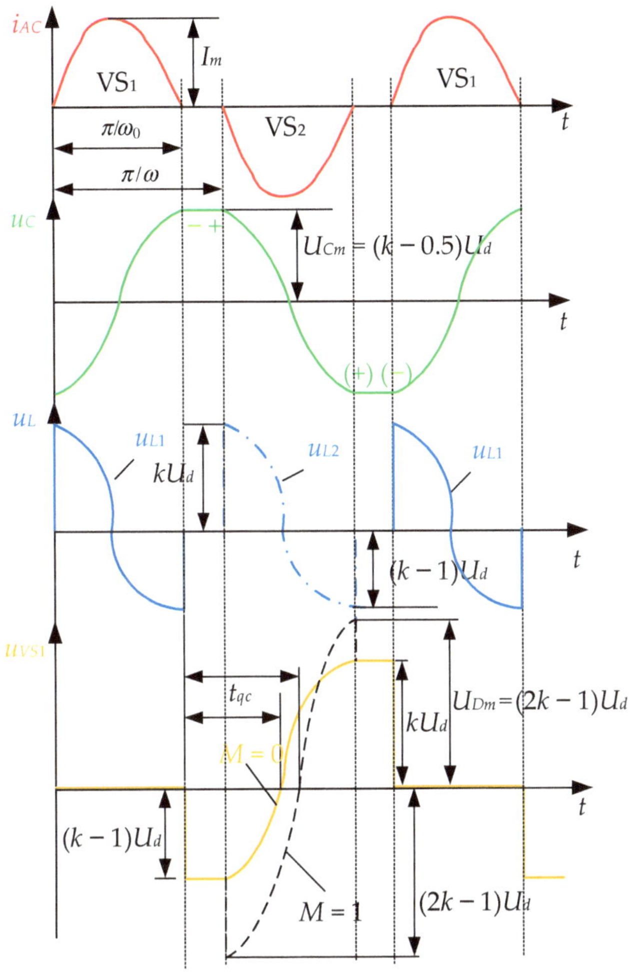

Figure 30. Timing diagrams describing the operating principle of a half-bridge thyristor RI. Source: Figure by author.

As a result of the current in the AC circuit, the capacitor C is charged resonantly to a voltage higher than that which ensures its flow, $U_d/2$ (at $C_{F1} = C_{F2}$), with a polarity without brackets. After the end of the current pulse, which has a duration of π/ω_0, the operating thyristor VS_1 is blocked, and the capacitor C is charged to its maximum voltage, U_{Cm}. The thyristor VS_1 is restored by the reverse voltage applied to it, which is the difference between U_{Cm} and $U_d/2$. After a certain pause, since the interrupted current mode is selected (Figure 30), a control pulse is supplied to the other thyristor VS_2. During its operation, a current flows with a direction opposite to that of the previous half-period of operation of the circuit, which recharges the capacitor C to a voltage U_{Cm} with a polarity shown in brackets. The midpoint of the power source is created using filter capacitors C_{F1} and C_{F2}, which also ensure its low internal resistance. To minimize their influence on the processes in the circuit, they are chosen with a value at least 10 times larger than the resonant capacitor C.

The considered series RLC circuit is described by the following differential equation, regarding the voltage across the resonant capacitor [33]:

$$LC\frac{d^2u_c}{dt^2} + RC\frac{du_c}{dt} + u_c = u_{SW}, \quad (36)$$

where u_{SW} is the input voltage supplied to the AC series circuit. It is formed by the series operation of the semiconductor switches (in this case thyristors) of the DC/AC converter with a switching (control) frequency $f = 1/T$.

The analysis is performed assuming the ideality of all circuit elements. When operating in a steady state and fulfilling the condition for the considered RLC circuit, the following expressions for the current through the resonant inductance and the voltage on the resonant capacitor are valid [33,36]:

$$\begin{gathered} i_L(t) = i_{AC}(t) = \frac{U_{DC}+U_C(0)}{\omega_0 L}e^{-\delta t}sin\omega_0 t, \\ u_C(t) = U_{DC} - (U_{DC}+U_C(0))e^{-\delta t}\left(\frac{\delta}{\omega_0}sin\omega_0 t + cos\omega_0 t\right), \end{gathered} \quad (37)$$

where $U_C(0)$ reflects the initial condition of the resonant capacitor voltage, and $U_{DC} = U_d/2$ is the DC supply voltage providing the current in the resonant circuit.

From the condition for periodicity of the voltage on the resonant capacitor $u_C\left(\frac{\pi}{\omega}\right) = -U_C(0)$, its initial value is determined: $U_C(0) = (2k - 1)U_d/2 = (k - 0.5)U_d$; this value is also maximum, as shown in Figure 30.

After substituting the initial condition in Equation (37), the current through the load and the capacitor is determined by the expression:

$$i_{AC}(t) = \frac{kU_d}{\omega_0 L}e^{-\delta t}\sin\omega_0 t. \quad (38)$$

Since the current in the AC circuit has a shape close to sinusoidal, it can be assumed that its maximum value occurs in the middle of the thyristor conduction interval:

$$I_{ACm} = i\left(\frac{\pi}{2\omega_0}\right) = \frac{kU_d}{\omega_0 L}e^{-\delta\frac{\pi}{2\omega_0}}. \quad (39)$$

When current flows across the resonant inductance L, a voltage is applied:

$$u_L(t) = L\frac{di(t)}{d(t)} = kU_d e^{-\delta t}\left(cos\omega_0 t - \frac{\delta}{\omega_0}sin\omega_0 t\right), \quad (40)$$

as the voltage at the moment of switching on the next thyristor is $u_L(0) = kU_d$, and at the end of its operation, respectively, $u_L\left(\frac{\pi}{\omega_0}\right) = -(k-1)U_d$.

In the absence of a magnetic connection between the two resonant inductances L_1 and L_2 ($M = 0$) for the maximum values of the forward and reverse voltages on the thyristors, respectively, we obtain $U_{Dm} = kU_d$ and $U_{Rm} = (k - 1)U_d$. If there is a magnetic connection between L_1 and L_2 ($M = 1$), the maximum values of the voltages on the devices increase and become equal in absolute value $U_{Dm} = U_{Rm} = (2k - 1)U_d$. The presence of a magnetic connection also affects the circuit recovery time of the thyristors, as in the absence of a connection the circuit time is shorter, as shown by the voltage diagram of the thyristor VS_1 in Figure 30. The interrupted current mode is more favorable from the point of view of the recovery of the thyristors, because the pause time is added to the circuit recovery time t_{qc}. The determination of t_{qc} is performed by compiling equations for the balance of voltages in the circuit during the operation of the corresponding thyristors, which are as follows: $U_d - u_L = 0$ (without magnetic coupling) or $U_d - 2u_L = 0$ (with magnetic coupling); this process leads to the numerical solution of transcendental equations. With sufficient accuracy for engineering practice, for values of the coefficient $k > 1.3$, the following expression can be used to determine t_{qc}:

$$t_{qc} = \frac{1}{\omega} arctg \frac{1}{2\nu}\left(\frac{\delta}{\omega_0} + \frac{\omega_0}{\delta}\right). \tag{41}$$

In this way, all the quantities necessary for designing the resonant DC/AC converter have been determined.

2.3.3. Operation of Resonant DC/AC Converter with Parallel Compensated Load

One of the typical applications of an RI is as a high-frequency power source for implementing various electrical technologies such as induction heating and melting of metals, and soldering [17,20]. A characteristic of these loads is that they are active-inductive with a very low power factor $cos\varphi_L < 0.2$ and can change significantly during the technological process. To compensate for the inductive nature of the load, parallel compensation is often used, whereby the RI load is a parallel load circuit consisting of R, L and C. Figure 31 shows the circuit of a parallel RI, which has the same topology as the parallel CSI, the difference being in the value of the input (L_d)/resonant inductance (L_R). In addition, when operating in the soft commutation mode (ZCS) of the semiconductor elements, the resonant inductance can be divided into two parts located in the DC (L_{R1}) and AC (L_{R2}) circuits of the DC/AC converter.

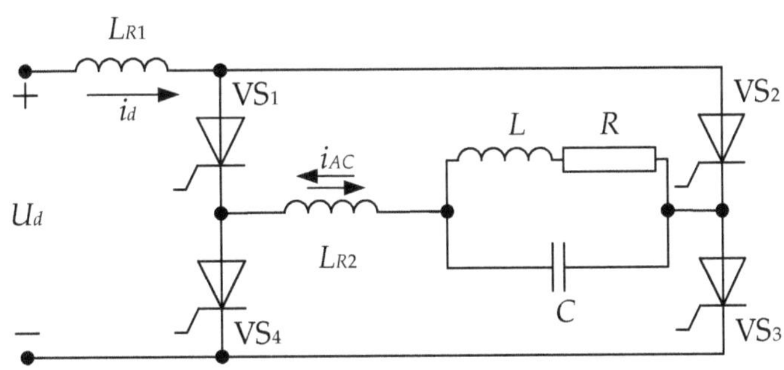

Figure 31. Full-bridge scheme of a thyristor parallel RI. Source: Figure by author.

In this topology, there are two resonant circuits, namely parallel (load) and series, composed of the resonant inductance and the equivalent parameters at the first harmonic of the switching frequency of the load circuit—$R_{(1)}$ and $X_{(1)}$. Analyses of parallel RIs have been presented in the literature, but unfortunately, due to the specifics of the power circuit (the presence of two branches in the circuit), the use of a series equivalent circuit is more appropriate. The equivalent circuit of the parallel RI is given in Figure 32. In order to ensure operation in the resonant mode of the series RLC circuit composed of L_R, $R_{(1)}$ and $X_{(1)}$, as well as the commutation of the thyristors, it is necessary for the parallel load circuit to operate in capacitive detuning, i.e., with $X_{(1)}$ having capacitive properties.

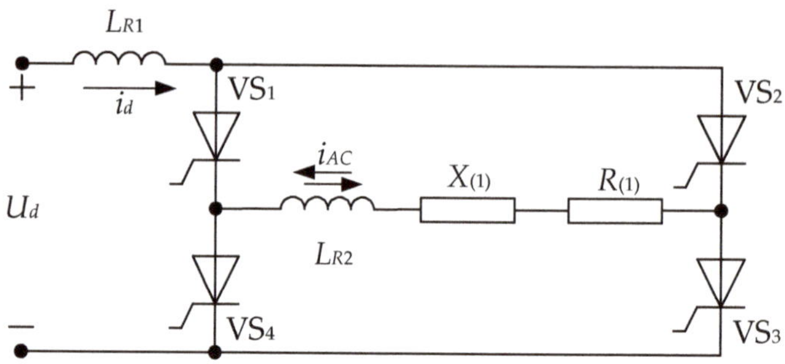

Figure 32. Series equivalent circuit of thyristor full-bridge parallel RI. Source: Figure by author.

In this context, for the use of a series equivalent circuit of a parallel RI, it is necessary to find the relationship between the values of the elements of the two resonant circuits. For this purpose, the vector diagram and the triangles of the conductances, presented in Figure 33, are used. For better clarity, the parallel equivalent circuit of the load from Figure 12 is used.

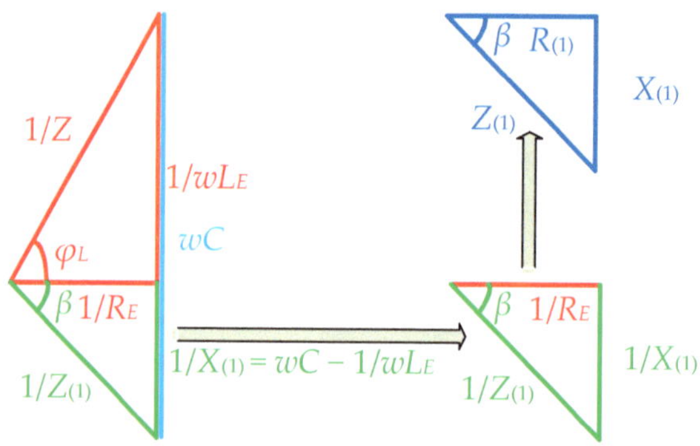

Figure 33. Equivalent circuits for determining the parameters of the serial equivalent circuit of the load circuit of a parallel RI. Source: Figure by author.

When operating in capacitive detuning of the parallel circuit, the load inductance from the parallel equivalent circuit of the load L_E is compensated by the parallel load capacitor C. As a result, a resultant first harmonic reactance of the load circuit is obtained $X_{(1)} = \omega C - 1/\omega L_E$, which is out of phase with the active component $R_{(1)}$ by the load circuit offset angle β. From this conductivity triangle (marked in green in Figure 33), the value of the first harmonic load impedance is determined:

$$Z_{(1)} = R_{(1)} cos\beta. \tag{42}$$

On the other hand, to switch to using a series equivalent circuit of the inverter AC circuit, the series equivalent circuit of the load circuit is used (shown in blue in Figure 33). Thus, the elements of the series equivalent circuit of the load circuit in the first harmonic are as follows:

$$R_{(1)} = R_E cos^2\beta \text{ and } X_{(1)} = R_E cos\beta sin\beta. \tag{43}$$

For low $cos\varphi_L$ loads, such as those that are common, the quality factor Q of the load circuit is above 3. In this case, it plays the role of a filter tuned to the frequency of the fundamental harmonic. All other harmonics are eliminated and their influence on the formation of the load voltage can be neglected. The analysis of the RI with a parallel load circuit is performed under the assumption that the circuit elements are ideal and active power is consumed only by the load. The DC power P_d supplying the inverter is equal to the active AC power delivered to the load P:

$$P_d = U_d I_d = P = I_{(1)} U cos\beta, \tag{44}$$

where U_d is the voltage of the DC power source; I_d is the consumed current of the DC power source; $I_{(1)}$ is the RMS value of the first harmonic of the RI current; and U is the RMS value of the load voltage. When operating in the BCM and a sufficiently large value of the coefficient k ($k > 1.3$), it can be considered that the inverter current is sinusoidal. In this case, the consumed current of the DC power source is obtained:

$$Id = \frac{1}{\pi}\int_0^{\pi} I_{(1)m} sin\theta d\theta = \frac{2}{\pi} I_{(1)m} = \frac{2\sqrt{2}}{\pi} I_{(1)} \tag{45}$$

Substituting into the power balance Equation (44) gives the following:

$$\frac{U}{U_d} = \frac{2\sqrt{2}}{\pi}\frac{1}{cos\beta} \tag{46}$$

which is the expression for the output characteristic of RI. When operating in the DCM, the RMS value of load voltage decreases with the increase in the pause and the harmonic composition of the load voltage worsens.

The division of the resonant inductance into the DC L_{R1} and AC L_{R2} parts of the circuit affects the voltages of the thyristors and their circuit recovery time. When fully included in the input circuit $L_R = L_{R1} + L_{R2}$, the longest circuit recovery time is provided, but also the highest voltages on the devices. The lowest voltages of the thyristors are obtained when dividing L_R into two equal parts in both DC and AC circuit.

2.3.4. Schematic Variations of Resonant DC/AC Converters

Resonant DC/AC Converters with Reverse Diodes

One way to improve the characteristics of the RI is to connect diodes with the reverse conductivity of each inverter semiconductor device in parallel. The circuit obtained in this way is known as the RI with reverse diodes (RIRD). A transistor bridge circuit of the RIRD with an active load is shown in Figure 34. A characteristic of the RIRD is that the voltage of the power devices is limited to the voltage of the DC power source U_d, regardless of changes in the load. Since during the conduction of the reverse diodes, energy is returned to the DC power source, for better use of the current devices, operation in a CCM is preferred. When operating with a control frequency lower than the resonant frequency of the RLC circuit (diagrams in Figure 35), after the operation of the next pair of devices, the voltage of the resonant capacitor C ensures the inclusion of their reverse diodes.

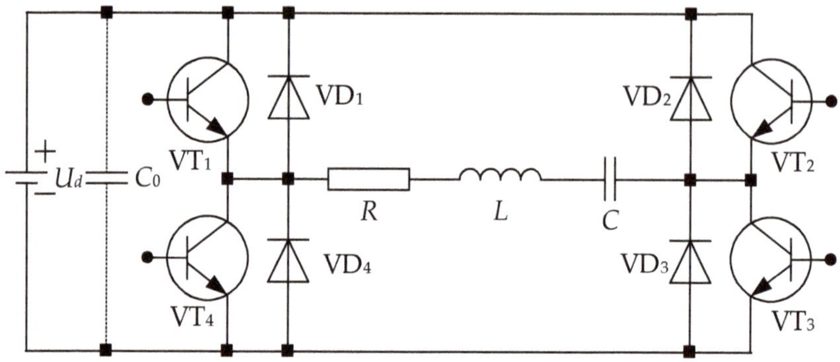

Figure 34. Full-bridge transistor RIRD. Source: Figure by author.

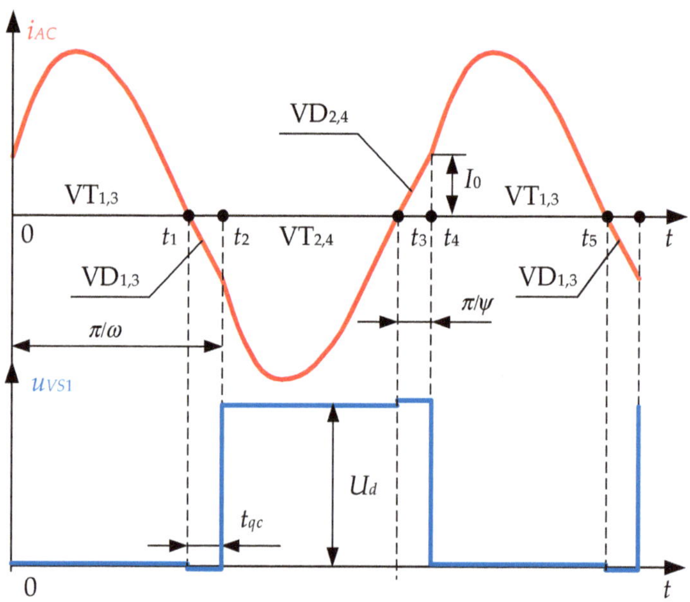

Figure 35. Full-bridge transistor RIRD, when operating with a control frequency below the resonant frequency. Source: Figure by author.

In this way, a reverse voltage is applied to the operated semiconductor switches. Therefore, in the operating mode with a frequency lower than the resonant frequency, single-operation thyristors can be used, and their switching off and recovery is realized by the voltage of the reverse diodes.

When operating in a mode with a control frequency higher than the resonant frequency of the RLC circuit (above the resonant frequency with diagrams shown in Figure 36), only transistors are used, and the current through them is terminated before it drops to zero.

The AC series RLC circuit of the RIRD is described by a differential equation regarding the capacitor voltage Equation (36). The equivalent power source u_{SW} is formed by the sequential operation of the semiconductor switches (transistors and reverse diodes) of the inverter with a switching frequency $f = 1/T$.

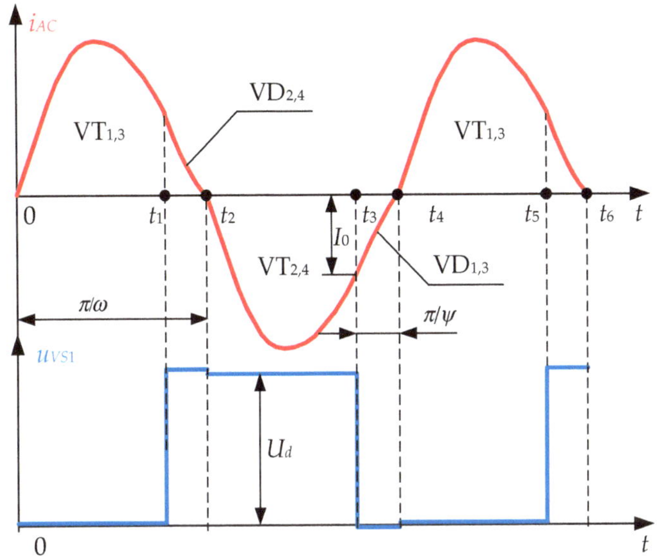

Figure 36. Full-bridge transistor RIRD, when operating with a control frequency above the resonant frequency. Source: Figure by author.

When operating in a steady state and fulfilling the resonance condition $R < 2\sqrt{\frac{L}{C}}$ for the considered circuit, the following equations for the current through the inductance and the capacitor voltage are valid [15,20,33]:

$$\begin{aligned} i_L(t) &= \frac{U_d + U_C(0)}{\omega_0 L} e^{-\delta t} sin\omega_0 t - I_L(0) e^{-\delta t} (cos\omega_0 t - \frac{\delta}{\omega_0} sin\omega_0 t) \\ u_C(t) &= U_d - (U_d + U_C(0)) e^{-\delta t} (\frac{\delta}{\omega_0} sin\omega_0 t + cos\omega_0 t) - \frac{I_L(0)}{\omega_0 C} e^{-\delta t} sin\omega_0 t \end{aligned} \tag{47}$$

where $I_L(0)$ and $U_C(0)$ reflect the initial conditions in the scheme.

In Figures 35 and 36, where the shape of the current through the inductance and the voltage on the semiconductor elements are given, when operating with a control frequency below the resonant and above the resonant frequency, the initial value of the current $I_L(0)$ and its initial phase ψ are marked.

From the analysis of the time diagrams, it is established that when operating the RIRD in an established mode, regardless of the ratio of the resonant and control frequencies, the conditions for periodicity of the state variables $i(\frac{\pi}{\omega}) = -I_L(0)$ and $u_C(\frac{\pi}{\omega}) = -U_C(0)$ are valid, and on this basis, by substitution in (47) their initial values are determined. On the other hand, the fact that the two state variables alternately change their signs at the beginning and at the end of the half-period of the control frequency (shown for the current through the inductance in the corresponding figures), allows us to obtain unified expressions for the mode with a control frequency below and above the resonant frequency. In this way, the values of the initial current and voltage are determined [37]:

$$I_L(0) = \frac{2KU_d}{\omega_0 L} a \text{ and } U_C(0) = (2K-1)U_d, \tag{48}$$

where $K = \frac{1}{1-h\cdot e^{\frac{-\delta\pi}{\omega}}} = \frac{1}{1-h\cdot\left(\frac{k-1}{k}\right)^{\frac{1}{v}}}$ is a quantity characterizing the series RLC circuit, called the oscillation coefficient in RIRD, and h and a are denoted as follows:

$$h = \frac{\frac{1}{\pi}ln\left(\frac{k}{k-1}\right)sin\frac{\pi}{v}+cos\frac{\pi}{v}+\left(\frac{k-1}{k}\right)^{\frac{1}{v}}}{\left(\frac{k-1}{k}\right)^{\frac{1}{v}}\left(\frac{1}{\pi}ln\left(\frac{k}{k-1}\right)sin\frac{\pi}{v}-cos\frac{\pi}{v}\right)-1},$$

$$a = \frac{sin\frac{\pi}{v}}{\frac{1}{\pi}ln\left(\frac{k}{k-1}\right)sin\frac{\pi}{v}-cos\frac{\pi}{v}-\left(\frac{k}{k-1}\right)^{\frac{1}{v}}}$$

After substituting the initial values of the state variables in (47), for the current in the AC circuit and the voltage on the resonant capacitor we obtain:

$$\begin{aligned} i_L(t) &= \frac{2KU_d}{\omega_0 L} D\, e^{-\delta t} sin(\omega_0 t - \psi') \\ u_C(t) &= U_d - 2KU_d E e^{-\delta t} sin(\omega_0 t + \varphi'), \end{aligned} \tag{49}$$

where $D = \sqrt{\left(1 + a\frac{\delta}{\omega_0}\right)^2 - a^2}$, $E = \sqrt{\left(\frac{\delta}{\omega_0} - a + a\left(\frac{\delta}{\omega_0}\right)^2\right)^2 - 1}$, $\psi' = arctg\frac{a}{1+a\frac{\delta}{\omega_0}}$ and $\varphi' = arctg\frac{1}{\frac{\delta}{\omega_0}-a+a\left(\frac{\delta}{\omega_0}\right)^2}$ are the initial phases of the current through the resonant inductance and the voltage across the resonant capacitor, respectively.

In the further analysis of the scheme, for greater generality of the considerations, it is necessary to normalize the conduction angle of the semiconductor switches relative to the control frequency ω, in which case the equations for $i_L(\theta)$ and $u_C(\theta)$ take the form:

$$\begin{aligned} i_L(\theta) &= \frac{2KU_d}{\omega_0 L} D\, e^{-\delta\frac{\theta}{\omega}} sin\frac{\pi}{\lambda}(\theta - \psi) \\ u_C(\theta) &= U_d - 2KU_d E\, e^{-\delta\frac{\theta}{\omega}} sin\frac{\pi}{\lambda}(\theta + \varphi), \end{aligned} \tag{50}$$

where $\theta = \omega t$, $\lambda = \pi\frac{\omega}{\omega_0}$, $\psi = \frac{\lambda}{\pi}\psi'$ and $\varphi = \frac{\lambda}{\pi}\varphi'$.

In order to apply a unified approach to analysis and to present the processes from a generalized perspective in all basic DC/AC converter circuits, it is appropriate to present the equation for the current in the AC circuit of the inverter from (50) as a function of the load resistance in the following way:

$$\begin{aligned} i_L(\theta) &= 4K\frac{U_d}{R}\frac{\delta}{\omega_0} D e^{-\delta\frac{\theta}{\omega}} sin\frac{\pi}{\lambda}(\theta - \psi) \\ u_C(\theta) &= U_d - 2KU_d E e^{-\delta\frac{\theta}{\omega}} sin\frac{\pi}{\lambda}(\theta + \varphi) \end{aligned} \tag{51}$$

Resonant DC/AC Converters with Voltage Limitation on the Resonant Capacitor (RI with Energy Dosing)

RIs are widely used in the process of converting electrical energy. This is due to some of their main advantages compared to other types of converters, namely the following [18,30,38,39]:

- Operation with electrical waveforms exhibiting a high degree of sinusoidal conformity, thereby minimizing harmonic distortion;

- Achievement of soft-switching conditions, resulting in inherently high efficiency without the incorporation of supplementary circuit elements;
- Effective exploitation of parasitic components as primary constituents in the formation of resonant circuits and the facilitation of resonant processes.

Conversely, the deployment of resonant converters is associated with a set of inherent limitations and challenges, including the following:

- The necessity of employing variable-frequency control strategies, which inherently increase the complexity of mitigating electromagnetic interference;
- A significant dependence of operational characteristics on the manufacturing tolerances of the components constituting the resonant network;
- Incomplete utilization of the semiconductor devices' rated power-handling capacity.

These challenges are typically addressed in two ways: either by improving the methods of synthesis of control of power electronic systems, including the use of artificial intelligence techniques [40–43] or by proposing power schemes that have a weak dependence of the operating modes on the load changes and have the ability to self-adjust the converter to the requirements of the specific application [16,44–47].

In this aspect, RIs with improved characteristics have been proposed, which eliminate the most significant part of the mentioned disadvantages, through the use of schemes with voltage limitation (fully or partially) on the resonant capacitor. These schemes have gained wide popularity in the specialized literature, as schemes with energy dosing (RIED). A number of RIED schemes are known [44], [46], with the single-stroke scheme having the lowest power, shown in Figure 37.

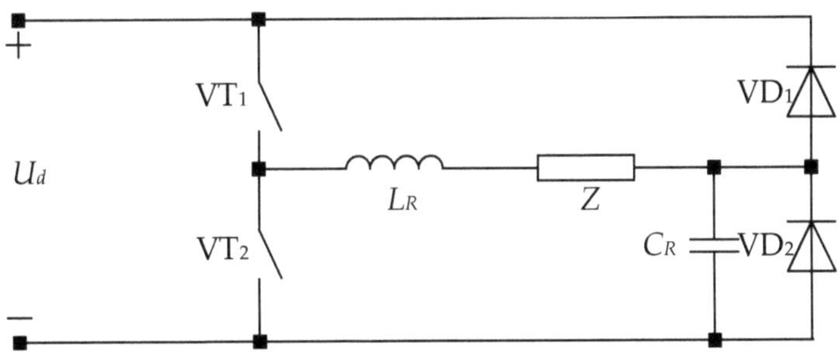

Figure 37. RIED single-stroke circuit. Source: Figure by author.

This DC/AC converter operates in a soft-switching mode, with diodes VD_1 and VD_2 limiting the voltage on the resonant capacitor in the range 0–U_d. This power circuit has found wide application in the implementation of electronic ballasts for powering fluorescent lamps. If the load is a parallel resonant circuit in capacitive detuning, the upper limit of the voltage on the resonant capacitor C_R can be changed within certain limits.

Among the various resonant converter topologies, the half-bridge configuration and the split resonant capacitor arrangement—depicted in Figure 38, are considered the most suitable for practical implementation. It is well established that the analytical treatment of each recognized topology can be tailored to a specific power stage through the introduction of numerical coefficients, thereby enabling the applicability of the derived results across different configurations.

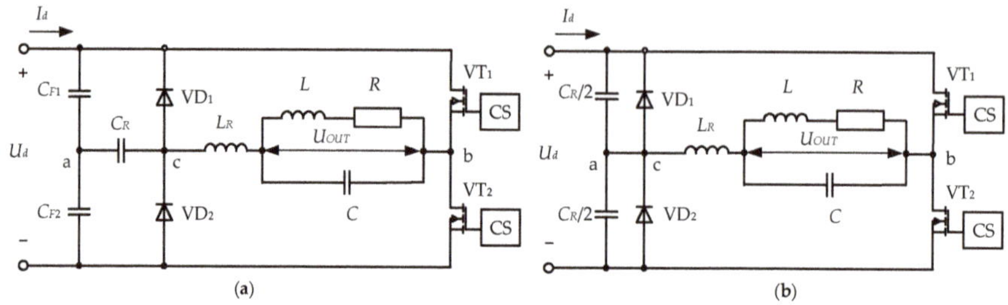

Figure 38. RIED circuit variants: (**a**) half-bridge; (**b**) with a split resonant capacitor. Source: Figure by author.

In practice, the split resonant capacitor topology is more frequently employed, as the prototyping of the half-bridge converter requires additional filter capacitors, which are absent in the former. The described power stage comprises two semiconductor switches VT_1 and VT_2, each equipped with an intrinsic anti-parallel diode, a resonant inductor L_R, and a resonant capacitor subdivided into two identical sections ($C_R/2$). The circuit further incorporates limiting (dosing) diodes VD_1 and VD_2, as well as a parallel load branch consisting of an inductor L (representing the load's inductive component), a resistor R (representing the load's resistive component), and a compensating capacitor C. The converter is energized from a constant-voltage DC source with output voltage U_d.

The choice of topology for clarifying the principle of operation of the RIED is not essential, because the circuit with a sectioned resonant capacitor is equivalent to the half-bridge circuit, provided that the ratio: $C_{R1} = C_{R2} = C_R/2$ is observed.

Figure 39 presents the timing diagrams of the current in the inverter's AC circuit and the voltage on the resonant capacitors, which visualize the electromagnetic processes in the RIED.

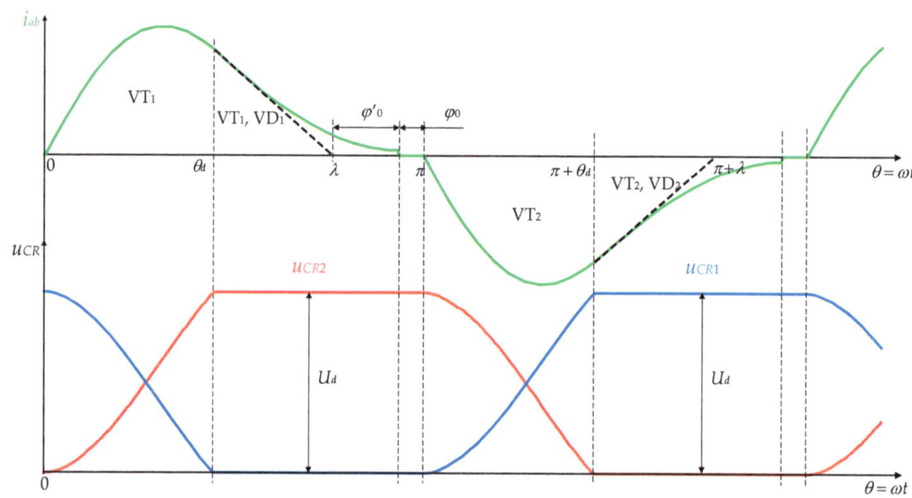

Figure 39. Timing diagrams in steady state operation of a RIED and a split resonant capacitor. Source: Figure by author.

From the type of diagrams in Figure 39, it can be concluded that within each half-period formed by the conduction of the corresponding transistor, three stages are observed—one of consuming energy from the power source (0–ϑ_d), one without consumption, which is characterized by a short circuit of the alternating current circuit (ϑ_d–($\pi - \varphi'_0$)), and a pause during which none of the semiconductor probes conduct—but in the parallel load circuit, resonant processes of energy exchange occur between the load inductance L and the parallel capacitor C.

An interesting feature of RIED is that the active power in the load is dosed, regardless of its changes. The energy of the resonant capacitor for one period of the control frequency is $W_{CR} = C_R U_d^2$ for half-bridge circuits and $W_{CR} = 4C_R U_d^2$ for full-bridge circuits. The amount of electricity that accumulates when charging the capacitor is, respectively,

$$\begin{aligned} Q_{CR} &= C_R U_d = 2I_d \tfrac{T}{2} \text{ for the half-bridge circuit;} \\ Q_{CR} &= C_R 2U_d = I_d \tfrac{T}{2} \text{ for the full-bridge circuit;} \end{aligned} \tag{52}$$

Therefore, the current consumed by the DC power source does not depend on the load parameters, but only on the control frequency f, and the expressions by which it is determined are the following:

$$\begin{aligned} I_d &= C_R U_d f \text{ for the half-bridge circuit;} \\ I_d &= 4\, C_R U_d f \text{ for the full-bridge circuit;} \end{aligned} \tag{53}$$

The expressions for the current through the load and the voltage on the resonant capacitor up to the time td in normalized form with respect to the control frequency ω are the same as for the classical RI [3,24] and are given by

$$\begin{aligned} i_{AC}(\vartheta) &= \tfrac{U_d}{\omega_0 L_R} e^{-\frac{\delta}{\omega}\vartheta} \sin\tfrac{\pi}{\lambda}\vartheta, \\ u_{CR}(\vartheta) &= \tfrac{U_d}{2} - \tfrac{U_d}{2}\sqrt{1 + \left(\tfrac{\delta}{\omega_0}\right)^2} e^{-\frac{\delta}{\omega}\vartheta} \sin\tfrac{\pi}{\lambda}(\vartheta + \phi). \end{aligned} \tag{54}$$

Using the fact that for angle ϑ_d the condition $U_{CR}(\vartheta_d) = U_d/2$ is fulfilled, then from the expression for the voltage of the C_R, we obtain the angle corresponding to the interval during which energy is consumed from the DC power source:

$$\vartheta_d = \left(\pi - \frac{\lambda}{\pi} arctg \frac{\omega_0}{\delta}\right). \tag{55}$$

The current in the load, after the moment t_d, corresponding to the angle ϑ_d (when the voltage across the entire resonant capacitor is limited and the load circuit is in resonance), is determined by the following equation:

$$i(\vartheta) = i(\vartheta_d) e^{-\delta\vartheta}. \tag{56}$$

Thus, power regulation within narrow limits is performed by changing the control frequency, and within wider limits—by using several values of the resonant capacitor, which can be switched by electronic switches.

Matching between the parameters of the inverter and the load can be realized by changing the value of the resonant capacitor or by dividing it into two parts, limiting the voltage of only one part. In these cases, the reverse voltage on the devices that are turned off is significant, which reduces the required circuit recovery time and, accordingly, allows for an increase in operating frequencies when using thyristors.

RIs are distinguished by a great variety of circuits. In the not-so-distant past, when thyristors were the main element base of power electronics, various ways were sought to increase the operating frequency of RI. One of the methods is the use of RIs with a doubling of the output frequency, in which the fundamental harmonic of the load voltage is twice-as-high a frequency as that with which the devices are controlled. Doubling is achieved by switching on the load at a point in the circuit where unidirectional current pulses flow, and their constant component is removed using an inverter transformer or a separation capacitor. In a doubling RI, an output current with a shape quite different from the sinusoidal one is obtained, and this method is used for consumers that are not sensitive to the content of harmonic components. Another method for increasing operating frequencies is multi-stage RI. They are built on the basis of different circuit variations of RI, and due to the fact that several single-stage inverters operate sequentially on a common load, a large pause occurs during the operation of each unit, which achieves a longer recovery time compared to single-stage circuits.

2.4. A Unified Approach for the Analysis of DC/AC Converters Based on the Study of Electromagnetic Processes in a Series RLC Circuit

2.4.1. Introduction

The continuous advancement in the characteristics and performance of power semiconductor devices—including higher voltage and current ratings, improved switching behavior, and lower losses—combined with the rapid development of sophisticated control systems, has driven the evolution of classical power converter topologies. These improvements have resulted in a significant enhancement of DC/AC converter capabilities, particularly in their operational frequency range, controllability, and overall efficiency.

Over the years, extensive research has been devoted to analyzing and designing the wide variety of DC/AC converter topologies available. This diversity stems from the equally diverse applications and load conditions with which these converters must be compatible. As a consequence, the need for a systematic approach to studying, comparing, and summarizing their operating principles and performance remains highly relevant.

Current-source inverters (CSIs) represent one of the fundamental categories of DC/AC converters. Their analysis most frequently employs the first harmonic approximation method [33,48–52]. In applications involving electrical technologies, where the load forms part of a resonant circuit with a sufficiently high quality factor, this approach delivers satisfactory accuracy when compared to exact analytical solutions that account for all harmonic components [1,2,5]. Its main advantage lies in its relatively low computational complexity, making it practical for many design tasks.

However, a specific characteristic of CSI-based systems is the requirement for a DC power supply of the "current source" type, without which they cannot operate. In most high-power applications, this source is realized by rectifying the AC mains through an AC/DC converter, followed by a filter inductor. Due to the properties of the supply network, this configuration behaves almost as an ideal DC voltage source with series inductance. Despite the usefulness of the first harmonic method, it exhibits certain limitations: it cannot provide a precise quantitative evaluation of the input current ripple or accurately determine the required inductance value. Furthermore, when the load is non-resonant or when higher-order harmonics cannot be neglected—such as in electric drives or certain energy applications—the method's accuracy becomes insufficient [3,21,29].

Voltage-source inverters (VSIs) form the second major category of DC/AC converters. Functionally, they are the dual counterpart of a CSI, being powered from a DC voltage source and generating a bipolar rectangular output voltage waveform (whereas CSI produces a bipolar rectangular output current). Here too, the first harmonic approximation is frequently applied [53–55], with limitations similar to those observed in CSI analysis. Since most VSI applications do not involve resonant loads, this method is generally restricted to preliminary estimations rather than final design calculations.

Resonant inverters (RIs) stand out by offering the greatest variety of topologies, operational modes, and application areas among DC/AC converters. They can be supplied either from a DC voltage source or, more rarely, from a DC current source. This flexibility broadens the range of possible circuit configurations, capabilities, and performance characteristics [18,30,39,56]. The classification proposed in [57] reduces resonant circuits to four principal types, with more complex matching networks typically being reducible to one of these basic configurations [15]. This systematic classification encompasses most practical RI designs.

In industrial applications, where simplicity, reliability, and cost-effectiveness are often primary considerations, the most widely adopted solutions are series resonant inverters (SRIs) powered from a DC voltage source. These configurations have been the focus of the most comprehensive research efforts [48–50,53]. Analytical approaches in this field often involve determining the actual current waveform in the AC circuit while neglecting the damping effect in the resonant network [1,5,6,13,19]. Such an approximation can yield acceptable results for resonant DC/DC converters; however, when applied to DC/AC resonant converters in industrial contexts—for example, induction heating systems—it may lead to significant deviations from the results obtained through exact analysis [16].

In contrast, accurate analytical methods account for multiple or all harmonic components of both the AC circuit current and the output voltage. These approaches provide a high degree of precision but are usually computationally intensive, complex to formalize, and thus difficult to integrate into automated design workflows. Their complexity also makes them less suitable for educational purposes and can hinder the clear interpretation of the obtained results.

The primary goal of this section is to present a unified analytical framework that applies to all major types of DC/AC converters. By describing the electromagnetic processes in these systems through a common set of parameters and calculation coefficients, it becomes possible to standardize and formalize their design methodologies. This unified approach is valuable both from a methodological perspective—particularly for power electronics education—and from a practical standpoint, as it facilitates the use of automated design tools, including those based on artificial intelligence techniques [41,43,58–60].

2.4.2. Basic Idea of the Unified Approach for Analyzing DC/AC Converters

For the analytical description of DC/AC converters operating in steady state, researchers in the field of power electronics employ a variety of approaches. The choice of method often depends on the type of converter, the nature of the load, and the desired level of accuracy in the calculations.

In the case of CSIs and VSIs, analysis typically involves the use of coefficients that incorporate both load parameters and control frequency. A common example is the load coefficient B, introduced in Section 2.1, which reflects the combined influence of the load characteristics and the converter's switching frequency. By employing such coefficients, the resulting analytical relationships clearly describe how variations in load or control frequency affect the operational behavior of the converter.

In resonant DC/AC converters, the electromagnetic processes in the AC circuit have a distinctly resonant nature. Therefore, their analysis requires the introduction of parameters that capture these resonant properties. Two of the most frequently used are the following:

- Quality factor Q—A dimensionless parameter that characterizes the sharpness of the resonance and the ratio of stored to dissipated energy per cycle;
- Detuning coefficient ν—Defined as the ratio between the control (switching) frequency and the resonant frequency of the AC circuit.

Some authors [16] prefer to use the detuning angle of the equivalent resonant circuit instead of ν, as this leads to more compact expressions in design formulas. This approach has been particularly useful for developing a unified analytical framework that can be applied to both RIs and CSIs, especially when a CSI is analyzed using the first harmonic approximation.

In practice, strictly resonant operation is rare. This is due to two main factors:

- Load variability—The parameters of the load often change during operation.
- Output power control—Adjusting the output power generally requires changing the control frequency, which shifts the system away from perfect resonance.

To address these limitations, it is more practical to use the coefficient k, which serves as an analogue of the quality factor Q but provides direct information about the maximum voltage across the resonant capacitor, regardless of the degree of detuning. This coefficient is particularly convenient because it yields compact analytical expressions for all quantities of interest in the study and design of RI.

Another important normalized parameter is the conduction angle normalization λ, defined as the conduction angle of the semiconductor switches normalized to the control frequency. This parameter offers significant advantages for evaluating the behavior of different DC/AC converter types under varying operating modes. Its inclusion in analysis and design simplifies control synthesis and tuning, enabling more efficient controller design [33,35,36].

The essence of this unified analytical methodology can be demonstrated through several representative DC/AC converter circuits, arranged in chronological order to reflect their historical development.

The first high-power controllable semiconductor devices were thyristors, and early DC/AC converters naturally adopted topologies suited to their operational characteristics. Figure 40a presents the full-bridge parallel thyristor CSI, designed for electrotechnological applications. In this topology, the AC circuit of the converter includes a parallel load network consisting of the following:

- An RL load, where the inductive element corresponds to the heating coil or inductor used in the process.
- A compensating capacitor, forming a parallel resonant load circuit.

At the input, a large series inductance (theoretically infinite) is connected in series with the DC power source. This inductance ensures that the converter operates from a current-source type supply.

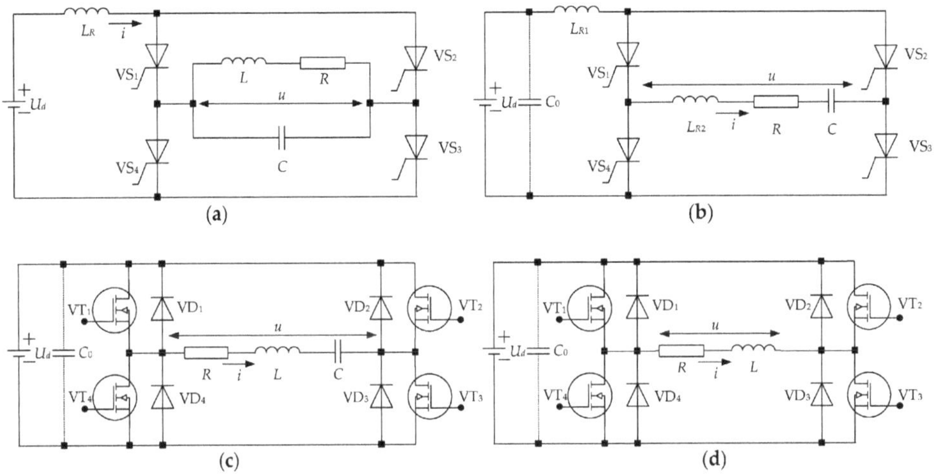

Figure 40. Full-bridge circuits of DC/AC converters: (**a**) parallel current-source inverter; (**b**) series resonant DC/AC converter; (**c**) resonant DC/AC converter with reverse diodes; (**d**) voltage source inverter. Source: Reprinted from [34], used with permission.

Due to the nature of thyristors, turn-off is achieved through the capacitive character of the parallel load circuit. The circuit is designed with capacitive detuning, ensuring natural commutation. In many RL-type induction heating processes, the load is predominantly inductive, meaning the parallel resonant circuit formed with the compensating capacitor has a high quality factor. This high-Q network also acts as an effective filter for the higher-order harmonics of both current and voltage.

For analysis using the first harmonic approximation, the load can be represented as a series connection of equivalent resistance and capacitance, simplifying calculations while preserving accuracy for high-Q conditions.

A variation of this topology, shown in Figure 40b employs a series-compensated load instead of a parallel one. This configuration offers different resonant characteristics and can be better suited to certain applications where load parameters or control objectives differ from those of the parallel-compensated arrangement.

To expand the range of possible operating modes in the power circuit, the resonant inductance is often split into two distinct components: L_{R1}, located in the DC input circuit, and L_{R2}, located in the AC circuit. This separation allows for greater flexibility in tuning the resonant characteristics and adapting the converter to different load conditions.

An important step in the evolution of resonant inverter topologies came with the introduction of high-power transistors into power electronic applications. This technological advancement led to the development of the transistor resonant inverter with reverse diodes (RIRD), illustrated in Figure 40c.

When the resonant capacitor is omitted from this configuration, the circuit's operational principle changes fundamentally, and the topology becomes equivalent to that of a VSI, as shown in Figure 40d.

If the load of the parallel CSI is assumed to be predominantly capacitive—a common scenario in many industrial applications—then, in combination with the input inductance (which is of very large inductance due to the requirements for creating an ideal current source), it forms a series resonant circuit. This circuit typically operates with a significant detuning between the resonant frequency and the control frequency—often by a factor of two to three or more.

Conversely, in the case of a voltage-source inverter, the load can be represented as an *R-L* network, which in the equivalent series RLC representation corresponds to a circuit where the capacitance value tends to infinity.

From these observations, a generalized theoretical conclusion emerges:

- The most universal structural model for DC/AC converters is the series RLC circuit;
- This circuit may operate in a resonant mode under certain conditions, or in a non-resonant (aperiodic) mode under others.

In this context, the resonant DC/AC converter represents the most general case, encompassing all possible operating modes. The other two main converter types—CSI and VSI—are therefore special cases derived from this general model.

This unifying viewpoint is visually summarized in Figure 41, which depicts the equivalent circuits of all major DC/AC converter types and the transitions between them.

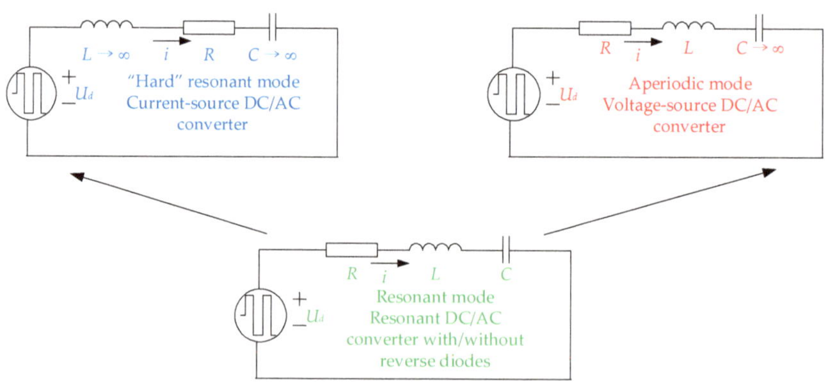

Figure 41. Generalized equivalent circuits of three main types of DC/AC converters. Source: Reprinted from [34], used with permission.

A common analytical feature of all these converters is the use of a series-equivalent AC circuit model in their analysis. This modeling approach enables the formulation of shared analytical relationships and allows a unified theoretical framework to be applied.

Accordingly, the unified approach can be expressed through the series RLC circuit, whose analysis describes the electromagnetic processes in the RIRD. The other converter types can be obtained as limiting cases of this model:

- CSI: Corresponds to the case where the resonant inductance tends to infinity.
- VSI: Corresponds to the case where the series capacitance tends to infinity, eliminating the possibility of resonant operation and resulting in aperiodic behavior.

When performing the analysis of resonant DC/AC converters, the damping factor k and the detuning factor of the equivalent AC circuit are applied. Also, for unification of operating modes, it is convenient to introduce a conduction angle of the semiconductor switches normalized to the control frequency $\lambda = \frac{\pi\omega}{\omega_0}$. When operating in aperiodic mode, by analogy with the resonant mode, the following parameters are introduced to be used in the analysis:

- aperiodicity coefficient $k_A = \frac{1}{1-e^{-\frac{\delta\pi}{\Omega_0}}}$;
- quasi-frequency coefficient $\nu_A = \frac{\omega}{\Omega_0}$, where $\Omega_0 = \sqrt{\delta^2 - \frac{1}{LC}}$ is a quasi-resonant frequency.

In this section, a unified analytical framework for the study of DC/AC converters was developed, with the resonant DC/AC converter with reverse diodes (RIRD) taken as the general case from which other topologies can be derived as special instances. Through the application of this generalized model—based on the equivalent series RLC circuit—explicit analytical expressions for the state variables were obtained, namely:

- The inverter (AC) current;
- The voltage across the resonant capacitor.

The derivations were initially performed for the full-bridge configuration, serving as the reference topology. To extend the applicability of the results, scaling coefficients were introduced, allowing the same analytical relationships to be directly adapted for alternative configurations, including the half-bridge, push–pull, and transformer midpoint topologies.

The strength of this approach lies in its methodological universality:

- It enables the use of a single mathematical model for different converter types.
- It simplifies the design process, reducing the need for topology-specific derivations.

It provides a foundation for automated design tools and facilitates the integration of artificial intelligence techniques in converter synthesis and optimization.

Ultimately, by formalizing the analysis in this way, the presented methodology ensures both educational clarity for training in power electronics and practical applicability in industrial converter design, regardless of the chosen topology or operating mode.

2.4.3. Analysis of a Resonant DC/AC Converter with Reverse Diodes

The analysis of the RIRD was discussed in detail in Section Resonant DC/AC Converters with Reverse Diodes, with the main emphasis placed on determining all the basic electrical quantities that characterize the operation of the circuit. The analysis defines all the key quantities necessary for the description and modeling of the inverter. Generalized expressions for the currents through the power semiconductor switches and the reverse diodes, as well as for the voltage on the resonant capacitor, are derived. The analysis provides an expression for the resonant frequency of the alternating current loop as a function of the resonant inductance and capacitance, taking into account the influence of the load and active resistances of the elements. On this basis, a design methodology is created in the following sections.

2.4.4. Analysis of a DC/AC Converter Operating in Hard Commutation Mode (Current-Source Inverter)

The CSI can be viewed as a specific case of a resonant inverter (RI) operating with a large ratio between the control frequency and the resonant frequency of the equivalent series AC circuit. In quantitative terms, this corresponds to detuning factor values greater than three, as established in previous studies [1,2,5,6,13,19]. For analytical and practical purposes, this operational state is referred to as the "hard" resonant mode. It stands in contrast to the "soft" resonant mode, where the disparity between the control and resonant frequencies is comparatively small—typically not exceeding 20–30%.

In industrial and electrotechnological applications, the parallel load circuit of the CSI (Figure 40a) most often operates in a condition of capacitive detuning. When considered at the fundamental harmonic, the equivalent parameters of this parallel load circuit, in combination with the input inductance, form a series resonant configuration. From a design perspective, the requirement for maintaining minimal input current ripple—essential for realizing a true current source power supply—necessitates that the input inductance be chosen with a very high value. This large inductance, in turn, results in a significantly reduced resonant frequency of the equivalent AC circuit.

The implications of this operating regime are best illustrated by examining the waveform patterns of the key state variables.

Figure 42 presents the current waveform in the AC circuit together with the corresponding voltage waveform across the equivalent capacitor when the converter is functioning in the "hard" resonant mode. These plots highlight the distinct phase relationships and amplitude behaviors characteristic of strong detuning, providing a visual basis for understanding the energy transfer dynamics and reactive power behavior in such systems.

In steady-state operation, the CSI maintains the same equivalent series AC circuit configuration as that of the RIRD. However, due to the substantially higher value of the input inductance in the CSI, the resonance condition $R < 2\sqrt{\frac{L}{C}}$ is satisfied even more rigorously for this topology. Consequently, the analytical expressions previously derived for the RIRD—namely, Equation (47) describing the inductor current and the capacitor voltage—remain directly applicable to the CSI as well.

A specific feature of this representation is that the parameters R and C in the series equivalent circuit do not correspond to physical discrete components but rather to equivalent values. These are determined according to the first harmonic approximation of the parallel load circuit, as given by Equation (43). In this context, R represents the equivalent active resistance accounting for real power dissipation, while C represents the equivalent reactive capacitance that models the reactive energy exchange in the system. This equivalent transformation greatly simplifies the analysis by reducing a complex parallel load network into a mathematically tractable series

RLC model, enabling the direct application of unified analytical methods to both CSI and RIRD topologies.

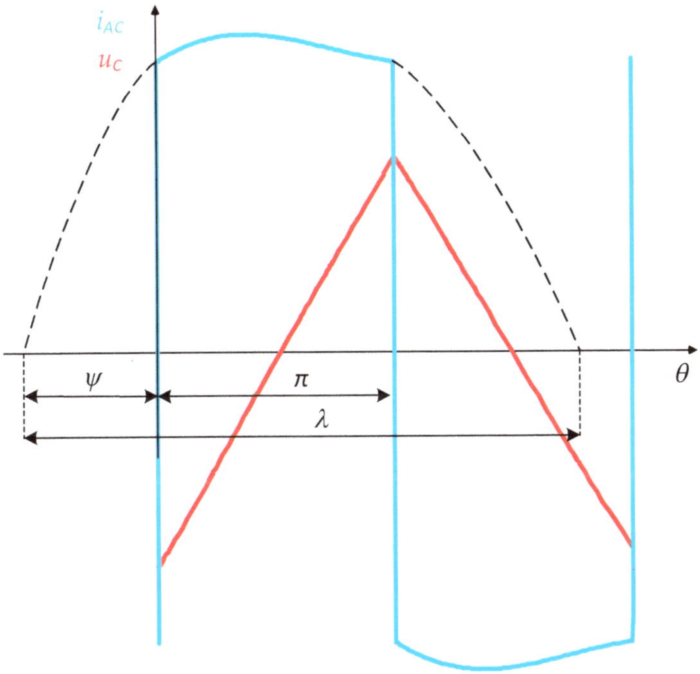

Figure 42. Shape of the current in the AC circuit and voltage across the equivalent series capacitor in a current-source inverter. Source: Reprinted from [34], used with permission.

The specificity of the operating mode is reflected by the periodicity conditions, which according to the timing diagrams in Figure 42 are as follows: $i_L\left(\frac{\pi}{\omega}\right) = I_L(0)$ and $u_C\left(\frac{\pi}{\omega}\right) = -U_C(0)$. After substitution in the expressions for the current in the AC circuit and the capacitor voltage, the following equations are obtained for determining the initial conditions of the state variables in the considered case:

$$I_L(0) = \frac{2K_{CS}\,U_d}{\omega_0 L}a \text{ and } U_C(0) = (2K_{CS} - 1)U_d, \tag{57}$$

where $K_{CS} = \frac{1}{1-h_{CS}\cdot e^{\frac{-\delta\pi}{\omega}}} = \frac{1}{1-h_{CS}\cdot\left(\frac{k-1}{k}\right)^{\frac{1}{v}}}$ is a quantity characterizing the series RLC circuit, called the oscillation coefficient in current inverters (hard commutation resonant inverters), and h_{CS} and a_{CS} are denoted as follows:

$$h_{CS} = \frac{\left(\frac{k-1}{k}\right)^{\frac{1}{v}} - \frac{1}{\pi}\,ln\left(\frac{k}{k-1}\right)sin\frac{\pi}{v} - cos\frac{\pi}{v}}{\left(\frac{k-1}{k}\right)^{\frac{1}{v}}\left(\frac{1}{\pi}ln\left(\frac{k}{k-1}\right)sin\frac{\pi}{v} - cos\frac{\pi}{v}\right) + 1},$$

$$a_{CS} = \frac{sin\frac{\pi}{v}}{\frac{1}{\pi}ln\left(\frac{k}{k-1}\right)sin\frac{\pi}{v} - cos\frac{\pi}{v} + \left(\frac{k}{k-1}\right)^{\frac{1}{v}}}.$$

After substituting the thus determined initial values of the state variables in (47), for the inverter current and the resonant capacitor voltage we obtain:

$$\begin{aligned} i_L(t) &= \tfrac{2K_{CS}\,U_d}{\omega_0 L} D_{CS}\, e^{-\delta t} sin(\omega_0 t + \psi'_{CS}), \\ u_C(t) &= U_d - 2K_{CS}\,U_d E_{CS} e^{\,-\delta t} sin(\omega_0 t + \varphi'_{CS}), \end{aligned} \tag{58}$$

where $D = \sqrt{\left(1 - a_{CS}\frac{\delta}{\omega_0}\right)^2 + a_{CS}^2}$, $E_{CS} = \sqrt{\left(\frac{\delta}{\omega_0} - a_{CS} - a_{CS}\left(\frac{\delta}{\omega_0}\right)^2\right)^2 + 1}$, $\psi'_{CS} = arctg\frac{a_{CS}}{1 - a_{CS}\frac{\delta}{\omega_0}}$ and $\varphi'_{CS} = arctg\frac{1}{\frac{\delta}{\omega_0} - a_{CS} - a_{CS}\left(\frac{\delta}{\omega_0}\right)^2}$ are the initial phases of the current through the resonant inductance and the voltage across the resonant capacitor, respectively.

In order to achieve greater universality of the considerations made, the conduction angle of the semiconductor switches was normalized with respect to the control frequency ω, whereby the expressions for $i_L(\theta)$ and $u_C(\theta)$ take the following form:

$$\begin{aligned} i_L(\theta) &= \tfrac{2K_{CS}\,U_d}{\omega_0 L} D_{CS}\, e^{-\delta\frac{\theta}{\omega}} sin\tfrac{\pi}{\lambda}(\theta + \psi_{CS}), \\ u_C(\theta) &= U_d - 2K_{CS}\,U_d E_{CS}\, e^{\,-\delta\frac{\theta}{\omega}} sin\tfrac{\pi}{\lambda}(\theta + \varphi_{CS}), \end{aligned} \tag{59}$$

where $\psi_{CS} = \frac{\lambda}{\pi}\psi'_{CS}$ and $\varphi_{CS} = \frac{\lambda}{\pi}\varphi'_{CS}$.

To implement a generalized analysis of the processes in all basic circuits of DC/AC converters, a transformation of the current expression from (59) has been performed, and it is presented as a function of the equivalent active load from the AC circuit:

$$\begin{aligned} i_L(\theta) &= 4K_{CS}\tfrac{U_d}{R}\tfrac{\delta}{\omega_0} D_{CS}\, e^{-\delta\frac{\theta}{\omega}} sin\tfrac{\pi}{\lambda}(\theta + \psi_{CS}), \\ u_C(\theta) &= U_d - 2K_{CS}\,U_d E_{CS}\, e^{-\delta\frac{\theta}{\omega}} sin\tfrac{\pi}{\lambda}(\theta + \varphi_{CS}), \end{aligned} \tag{60}$$

One of the significant advantages of treating the current-source inverter (CSI) as a resonant, hard-commutated inverter lies in the possibility of designing it with a finite input inductance, rather than assuming an idealized infinitely large value. This perspective not only broadens the practical applicability of CSI designs but also enables optimization of the input stage for improved size, cost, and dynamic performance.

The determination of the optimal value of the input inductance becomes a fundamental design task, directly affecting current ripple, energy storage requirements, and the overall efficiency of the converter. Achieving this balance ensures that the input inductance is sufficiently large to approximate current-source behavior while avoiding unnecessary oversizing, which would increase volume, weight and cost. This problem—often referred to as the optimal input inductance determination—is central to the design methodology of CSI systems and has been extensively addressed in the literature [3,4,33,36,61].

2.4.5. Analysis of a Resonant DC/AC Converter Without Reverse Diodes

It is well established that resonant DC/AC converters without reverse diodes are capable of operating in several distinct modes. In contrast to the previously discussed current-source inverter topology, these converters typically operate with a control frequency that is lower than, or at most equal to, the resonant frequency of the AC circuit.

For analytical modeling, the same equivalent series RLC circuit described in the earlier case remains valid. The essential difference, however, lies in the initial conditions: at the beginning of each switching interval, the current through the inductance is zero. This fundamental change in the initial state simplifies the analytical derivation, allowing for more concise mathematical expressions for the state variables compared to the general solution presented in Equation (49):

$$\begin{gathered} i_L(t) = \frac{U_d + U_C(0)}{\omega_0 L} e^{-\delta t} sin\omega_0 t, \\ u_C(t) = U_d - (U_d + U_C(0)) e^{-\delta t} \left(\frac{\delta}{\omega_0} sin\omega_0 t + cos\omega_0 t\right), \end{gathered} \tag{61}$$

From a design perspective, this simplification has practical value. It enables engineers to perform faster preliminary calculations, facilitates the optimization of component values for desired operating points, and aids in the tuning of the control system to maintain efficiency across varying load conditions.

Moreover, the reduced complexity of the analytical model makes it more suitable for educational purposes, for rapid prototyping, and for integration into automated design tools, including those based on artificial intelligence-assisted parameter optimization.

For the voltage on the equivalent capacitor, the condition for periodicity in the steady state is valid: $u_C\left(\frac{\pi}{\omega}\right) = -U_C(0)$. After substitution in the expression for the capacitor voltage, for its initial condition we obtain:

$$U_C(0) = (2k - 1)U_d \tag{62}$$

Equation (62) defines the oscillation coefficient. After substituting the initial condition for the capacitor voltage, Equation (61) take the form

$$\begin{gathered} i_L(t) = \frac{2kU_d}{\omega_0 L} e^{-\delta t} sin\omega_0 t, \\ u_C(t) = U_d - 2kU_d e^{-\delta t} \left(\frac{\delta}{\omega_0} sin\omega_0 t + cos\omega_0 t\right), \end{gathered} \tag{63}$$

If a compact description is introduced for the capacitor voltage, then the state variables are defined as follows:

$$\begin{gathered} i_L(t) = \frac{2kU_d}{\omega_0 L} e^{-\delta t} sin\omega_0 t, \\ u_C(t) = U_d - 2kU_d E_R e^{-\delta t} sin(\omega_0 t + \varphi'_R), \end{gathered} \tag{64}$$

where $E_R = \sqrt{1 + \left(\frac{\delta}{\omega_0}\right)^2}$, $\varphi'_R = arctg\frac{\omega_0}{\delta}$ is the initial phase of the voltage across the resonant capacitor.

After normalizing the conduction angle of the semiconductor switches with respect to the control frequency ω, the expressions for $i_L(\theta)$ and $u_C(\theta)$ are:

$$\begin{aligned} i_L(\theta) &= \tfrac{2kU_d}{\omega_0 L} k e^{-\delta\frac{\theta}{\omega}} sin\tfrac{\pi}{\lambda}\theta, \\ u_C(t) &= U_d - 2kU_d E_R\, e^{-\delta\frac{\theta}{\omega}} sin(\omega_0 t + \varphi_R), \end{aligned} \tag{65}$$

where $\varphi_R = \frac{\lambda}{\pi}\varphi'_R$.

Accordingly, after expressing the state variables as functions of the equivalent load, we obtain:

$$\begin{aligned} i_L(\theta) &= 4k\tfrac{U_d}{R}\tfrac{\delta}{\omega_0} k\, e^{-\delta\frac{\theta}{\omega}} sin\tfrac{\pi}{\lambda}\theta \\ u_C(t) &= U_d - 2kU_d E_R e^{-\,\delta\frac{\theta}{\omega}} sin(\omega_0 t + \varphi_R) \end{aligned} \tag{66}$$

Using Equation (66), all the quantities necessary for the design of the RI are determined.

2.4.6. Analysis of an RLC Inverter Operating in Aperiodic Mode

To ensure operation of the RLC inverter in aperiodic mode, the condition must be met $R > 2\sqrt{\frac{L}{C}}$. In this case, the state variables are defined by:

$$\begin{aligned} i_L(t) &= \tfrac{U_d + U_C(0)}{\Omega_0 L} e^{-\delta t} sh\Omega_0 t - I_L(0) e^{-\delta t}(ch\Omega_0 t - \tfrac{\delta}{\Omega_0} sh\Omega_0 t), \\ u_C(t) &= U_d - (U_d + U_C(0)) e^{-\delta t}(\tfrac{\delta}{\Omega_0} sh\Omega_0 t + ch\Omega_0 t) - \tfrac{I_L(0)}{\Omega_0 C} e^{-\delta t} sh\Omega_0 t \end{aligned} \tag{67}$$

From the joint solution of Equation (67), using the conditions for periodicity of the current through the load and the voltage on the capacitor, i.e., $i_L\left(\frac{\pi}{\omega}\right) = I_L(0)$ and $u_C\left(\frac{\pi}{\omega}\right) = U_C(0)$—the values of the initial current and voltage are found, namely $I_L(0) = \frac{2K_{ap}U_d}{\Omega_0 L} a_{ap}$ and $U_C(0) = (2K_{ap} - 1)U_d$, where $K_{ap} = \frac{1}{1-h_{ap}e^{\frac{-\delta\pi}{\Omega_0}}} = \frac{1}{1-h_{ap}\left(\frac{k_A-1}{k_A}\right)^{\frac{1}{\nu_A}}}$ is a quantity characterizing the series RLC circuit, called the aperiodicity coefficient, and h_{ap} and a_{ap} are denoted by:

$$\begin{aligned} h_{ap} &= \frac{-\frac{1}{\pi}ln\left(\frac{k_A}{k_A-1}\right) sh\frac{\pi}{\nu_A} - ch\frac{\pi}{\nu_A} - \left(\frac{k_A-1}{k_A}\right)^{\frac{1}{\nu_A}}}{1+\left(\frac{k_A-1}{k_A}\right)^{\frac{1}{\nu_A}}\left(ch\frac{\pi}{\nu_A} - \frac{1}{\pi}ln\left(\frac{k_A}{k_A-1}\right) sh\frac{\pi}{\nu_A}\right)}, \\ a_{ap} &= \frac{sh\frac{\pi}{\nu_A}}{\left(\frac{k_A}{k_A-1}\right)^{\frac{1}{\nu_A}} + ch\frac{\pi}{\nu_A} - \frac{1}{\pi}ln\left(\frac{k_A}{k_A-1}\right) sh\frac{\pi}{\nu_A}}. \end{aligned}$$

After substituting the initial current and voltage into Equation (67), the following expressions for the inductor current and capacitor voltage are obtained:

$$\begin{aligned} i_L(t) &= \tfrac{2K_{ap}\,U_d}{\Omega_0 L} D_A\, e^{-\delta t} sh(\Omega_0 t - \psi'_A), \\ u_C(t) &= U_d - 2K_{ap}U_d E_A e^{-\delta t} sh(\Omega_0 t + \phi'_A) \end{aligned} \tag{68}$$

where $D_A = \sqrt{\left(1+a_{ap}\frac{\delta}{\Omega_0}\right)^2 - a_{ap}^2}$, $\psi'_A = arth\frac{a_{ap}}{1+a_{ap}\frac{\delta}{\Omega_0}}$, $E_A = \sqrt{\left(\frac{\delta}{\Omega_0} - a_{ap} + a_{ap}\left(\frac{\delta}{\Omega_0}\right)^2\right)^2 - 1}$ and $\varphi'_A = arth\frac{1}{\frac{\delta}{\Omega_0} - a_{ap} + a_{ap}\left(\frac{\delta}{\Omega_0}\right)^2}$.

After normalizing the conduction angle of the semiconductor switches with respect to the control frequency, the expressions for the inductor current and the capacitor voltage are obtained as follows:

$$\begin{aligned} i_L(\theta) &= \frac{2K_{ap}\,U_d}{\Omega_0 L} D_A\, e^{-\delta\frac{\theta}{\omega}} sh\frac{\pi}{\lambda_A}(\theta - \psi_A), \\ u_C(\theta) &= U_d - 2K_{ap}U_d E_A e^{-\delta\frac{\theta}{\omega}} sh\frac{\pi}{\lambda_A}(\theta + \phi_A) \end{aligned} \tag{69}$$

where $\theta = \omega t$, $\lambda_A = \frac{\pi\omega}{\Omega_0}$, $\psi_A = \frac{\lambda_A}{\pi}\psi'_A$ and $\varphi_A = \frac{\lambda_A}{\pi}\varphi'_A$.

2.4.7. Voltage-Source Inverter Analysis

The foregoing considerations regarding the aperiodic operation of a series RLC inverter form the basis for analyzing the VSI. The innovative aspect of this approach lies in treating the VSI as a special case of a series RLC inverter with reverse diodes, operating in aperiodic mode when the capacitance value is $C = \infty$ [55]. This interpretation enables the direct application of the previously derived relationships for VSI analysis, with the understanding that for $C = \infty$, the detuning coefficient becomes $\delta = \Omega_0$.

In the analysis of a VSI, the aperiodicity coefficient acquires a single fixed value $k_A = k_N = \frac{1}{1-e^{-\pi}} = 1.045$.

Then the expression for the current through the inductance, represented by its initial phase, is the following:

$$i_L(t) = \frac{U_d}{\Omega_0 L} D_N e^{-\delta t} sh(\Omega_0 t - \psi'_N) = \frac{2U_d}{R} D_N e^{-\delta t} sh(\Omega_0 t - \psi'_N), \tag{70}$$

where $D_N = \sqrt{(1+a_N)^2 - a_N^2}$, $\psi'_N = arth\frac{a_N}{1+a_N}$, $a_N = \frac{sh\frac{\pi}{\nu_A}}{\left(\frac{k_N}{k_N-1}\right)^{\frac{1}{\nu_A}} + ch\frac{\pi}{\nu_A} - sh\frac{\pi}{\nu_A}}$.

By normalizing with respect to the control frequency ω, the expression for $i_L(\theta)$ takes the form

$$i_L(\theta) = \frac{U_d}{\Omega_0 L} D_N\, e^{-\delta\frac{\theta}{\omega}} sh\frac{\pi}{\lambda_A}(\theta - \psi_N) = \frac{2U_d}{R} D_N\, e^{-\delta\frac{\theta}{\omega}} sh\frac{\pi}{\lambda_A}(\theta - \psi_N), \tag{71}$$

where $\theta = \omega t$, $\lambda_A = \frac{\pi\omega}{\Omega_0}$ and $\psi_N = \frac{\lambda_A}{\pi}\psi'_N$.

In this mode, the maximum value of the load current occurs at the angle $\theta_m = \pi$ and is:

$$I_{\max} = i_L(\theta_m) = i_L(\pi) = \frac{2U_d}{R} D_N\, e^{-\delta\frac{\theta_m}{\omega}} sh\frac{\pi}{\lambda_A}(\theta_m - \psi_N) \tag{72}$$

The average value of the current through the transistors is determined by solving the integral:

$$I_{av} = \frac{1}{2\pi}\int_{\psi_N}^{\pi} i_L(\theta)d\theta = \frac{1}{2\pi}\int_{\psi_A}^{\pi} \frac{2U_d}{R} D_N\, e^{-\delta\frac{\theta}{\omega}} sh\frac{\pi}{\lambda_A}(\theta-\psi_N)d\theta \tag{73}$$

and after the transformations it is

$$I_{av} = \frac{U_d D_N}{\pi R F_A}\left(e^{-\frac{\delta}{\omega}\pi} ch\left(\alpha + \frac{\pi}{\lambda_A}(\pi-\psi_N)\right) - e^{-\frac{\delta}{\omega}\psi_N} ch\alpha\right), \tag{74}$$

where, $F_A = \sqrt{\left(\frac{\pi}{\lambda_A}\right)^2 - \left(\frac{\delta}{\omega}\right)^2}$ and $\alpha = arth\frac{\frac{\delta}{\omega}}{\frac{\pi}{\lambda_A}}$.

The average value of the current consumed by the DC power source is determined by the expression:

$$I_d = \frac{1}{\pi}\int_{0}^{\pi} i_L(\theta)d\vartheta = \frac{1}{\pi}\int_{0}^{\pi} \frac{2U_d}{R} D_N\, e^{-\delta\frac{\theta}{\omega}} sh\frac{\pi}{\lambda_A}(\theta-\psi_N)d\theta \tag{75}$$

and after its solution it is

$$I_d = \frac{2U_d D_N}{\pi R F_A}\left(e^{-\frac{\delta\pi}{\omega}} ch\left(\alpha + \frac{\pi}{\lambda_A}(\pi-\psi_N)\right) - ch\left(\alpha - \frac{\pi}{\lambda_A}\psi_N\right)\right) \tag{76}$$

The average current through the diodes is determined by the expression:

$$I_{dav} = I_{av} - \frac{I_d}{2} \tag{77}$$

The RMS value of the load current is

$$I = \sqrt{\frac{1}{\pi}\int_{0}^{\pi} i_L^2(\theta)d\vartheta} = \sqrt{\frac{1}{\pi}\int_{0}^{\pi}\left(\frac{2U_d}{R} D_N\, e^{-\delta\frac{\theta}{\omega}} sh\frac{\pi}{\lambda_A}(\theta-\psi_N)\right)^2 d\theta} = \frac{U_d D_N}{R} G_N \tag{78}$$

where

$G_N = \sqrt{\frac{1}{\pi}\left(\frac{1}{F_A}\left(e^{-2\pi\frac{\delta}{\omega}} sh\left(\alpha + \frac{2\pi}{\lambda_A}(\pi-\psi_N)\right) - sh\left(\alpha - \frac{2\pi}{\lambda_A}\psi_N\right)\right) + \frac{\omega}{\delta}\left(e^{-2\pi\frac{\delta}{\omega}} - 1\right)\right)}$.

For the RMS value of the voltage on the active part of the load R, we obtain:

$$U = IR = U_d D_N G_N. \tag{79}$$

On the other hand, the RMS value of the first harmonic of the inverter voltage is determined by

$$U = \frac{2\sqrt{2}}{\pi} U_d = \frac{U_d D_N}{R} G_N \tag{80}$$

It should be emphasized that all the derived relationships are expressed as functions of the aperiodicity coefficients (for the considered fixed-value operating regime) and the quasi-frequency coefficient. These expressions define all the fundamental quantities required for the design of voltage inverters. Consequently, they provide the basis for determining the loading conditions of all circuit elements and for accurately sizing them to meet design requirements.

2.4.8. Conclusion and Findings

The presented analysis establishes a unified methodological framework for studying the three principal classes of DC/AC converters, namely resonant inverters with reverse diodes, current-source inverters, and voltage-source inverters, as special cases of a generalized series RLC topology. By describing the AC circuit current in its real form and incorporating the active resistance of the circuit, the approach ensures higher accuracy than conventional methods based solely on idealized assumptions or simplified harmonic models.

One of the essential benefits of this unified treatment is the ability to directly compare the characteristics of different inverter types within a single analytical framework. This not only supports the development of standardized design procedures but also allows for seamless adaptation of results from one topology to another through the use of numerical coefficients. Furthermore, the method enables precise evaluation of parameter variations and their impact on performance, which is critical for both optimization and practical implementation.

Despite its broad applicability, the proposed approach does not yet extend to grid-tied converters, which play a vital role in modern distributed generation systems. Future work should focus on expanding the methodology to cover these topologies, integrating the influence of grid dynamics and compliance with power quality standards.

The combination of the unified inverter model with artificial intelligence-based optimization techniques represents a promising avenue for advanced design automation. Optimization algorithms—whether evolutionary, gradient-based, or hybrid—can exploit the compact analytical expressions derived here to rapidly converge on optimal component values and control settings.

From an educational perspective, this framework also enhances the teaching of power electronics. The normalization of the semiconductor conduction angle with respect to control frequency offers a clear, comparable basis for analyzing different modes of operation, while considering the current-source inverter as a hard-commutated resonant system provides valuable insight into the design of devices with finite input inductance—addressing a key challenge in practice.

In conclusion, the unified analytical approach not only refines inverter theory but also paves the way toward intelligent, optimization-driven design methodologies. Its adoption can significantly improve both the efficiency of the design process and the performance of the resulting systems, ensuring that the next generation of DC/AC converters meets the increasingly stringent demands of modern industrial, renewable energy, and educational applications.

3. Modeling of DC/AC Converters

3.1. Introduction

In modern conditions, the requirements for electrical energy converters are constantly increasing. To achieve better indicators such as: high efficiency, reduced dimensions, weight and price, as well as improved stability and reliability during operation, new methods and tools for automated design are increasingly used. These modern tools also lead to the improvement of the parameters characterizing the quality of the output or consumed electrical energy, and improve the methods of control and protection of devices. Increasing the productivity of design engineers, a significant reduction in material costs and shortening the development and implementation times are realized through the use of modern achievements in mathematical modeling, computational mathematics, computer technology and software engineering. The strong development of computer technology has led to significant progress in other areas of science and technology. The role of human creative activity in the "research–production" cycle has significantly increased. There are three main groups of methods for conducting research, including in the field of power electronics [15,53,62–64]:

1. Analytical methods: Analytical solutions are based on a number of simplifying assumptions and limitations that can reduce the accuracy of the study. In complex objects, the application of analytical methods is often associated with a large amount of computational work.

2. Experimental methods: Experimentation is an important criterion for the correct solution of a number of research tasks. However, the practical application of experimental methods is associated with a number of problems, such as the need to create a real physical object and a suitable external environment for its existence, which is not always easy to achieve. Finding errors in experiments is difficult, and the study of intermediate, transient and emergency states is not always possible.

3. Modeling methods: The advantages of this method are numerous: great opportunities for research, universality and the possibility of repeated use of models in various developments; the study can be conducted before the implementation of the object; a real external environment is not needed, since it can be modeled; the ability to study boundary situations, intermediate, transient and emergency regimes that can be dangerous for the real system or last only for a very short time; the ability to scale processes; this method is the only one for studying complex systems and is used to create training systems.

A model is a system that, whether represented conceptually or materially realized, reflects and reproduces the object of study. It has the ability to replace it in such a way that its study provides information about this object [15].

The main characteristics of the model are as follows: The model always reflects or reproduces a certain aspect of reality. The purpose of creating the model is cognitive—to obtain new information about the object of study. The model represents a specific system that is accessible for observation, research or practical actions. For each model, there are clear rules that define how to move from information about the model itself to information about the object that it represents.

In this context, modeling is a method of scientific knowledge that combines three main types of activities: first, building a model of the object under study; second, conducting "experiments" with the model; third, transforming the results of these experiments into information about the object of study itself.

Modeling, performed using computers (digital computers) is called computer (digital) modeling, and models implemented using computer technology are known as computer (digital) models. Computer models are mathematically similar models, in which the similarity between the model and the object is based mainly on isofunctionalism, and less often on isomorphism. Computer modeling is a powerful technique with extremely wide application in all areas of research. The tendency to expand the use of computer models is determined by a number of advantages that they offer [65–67]:

- Computer models provide the opportunity to make decisions already at the stage of developing future systems, which significantly reduces the need for costly "trial and error" methods. The parameters of the modeled system can be studied under a wide range of changes in its influencing factors, including those that cannot be implemented or controlled in real conditions. For example, the computer model can be studied with input data almost without restrictions, while for real systems certain combinations of factors can be difficult to implement or dangerous for the system itself.
- The results of the real operation of the modeled object can be extrapolated onto the computer model, the goal being to predict its behavior. Furthermore, computer modeling allows processes to evolve many times faster than in real systems, which is particularly useful when studying fast processes or when scaling up over time.
- Random elements related to the system under study can be included in the model, which allows conclusions to be drawn about the system in a certain probabilistic context. Digital modeling is often the only method applicable to the study and improvement of complex systems and in many cases is the most economical method. In addition, this method provides opportunities for training and self-education of specialists, including in the field of power electronics.

The advantages provided show that computer modeling is an extremely powerful tool for research. However, its use should not become an end in itself. The effectiveness of models depends entirely on their correct application. There are cases when computer modeling is fully justified and even the only possible, but there are also situations in which the use of other methods may be more effective and convenient. In order to make the right decision about the appropriateness of computer modeling in a given case, it is necessary to know the object of research and the modeling technique well. In addition, criteria for the application of the models must be formulated, and in many cases the decision also depends on other factors, such as the specificity of the problem, the timeliness of the results, and the available resources.

3.2. Modeling Stages and Adequacy of Computer Models Used in Power Electronics

The process of modeling through computer (digital) modeling includes several consecutive stages, which begin with the creation of the program model and end with the analysis and evaluation of the results of the experiments [15,65–67]. Initially, the construction of the computer model includes three substages: creating the concept of the model, creating a model scheme and programming the model. Each of these substages is important for checking the reliability of the model. At the stage of creating the concept of the model, the idea of the model is formulated, conclusions are drawn, decisions are made and important aspects such as methodology, resources and division of the task into subtasks are considered. In this substage, real processes are studied and approximations and assumptions that will be used in the model are determined. The conceptual model is a description of the model in abstract terms and concepts, clarifying the purpose, functions, approximations and operating modes. In the second substage, the logical block diagram of the model is created, which concretizes the abstract idea from the conceptual model. This stage often uses the modular principle, with the scheme representing a collection of separate blocks that are interconnected. At the stage of programming the model, the scheme is written using an appropriate programming language, which is the next step towards the implementation of the model.

After the construction of the program model, the stage of experimentation with the model follows. This stage includes two substages: planning xperiments and conducting experiments. When planning experiments, the values of the parameters and their combinations are defined, taking into account the resources of the computer. When conducting experiments, the input information is prepared, calculations are performed using the computer and the results are obtained. These experiments can be control or working. In control modeling, the operability of the model is checked and the sensitivity of the results to changes in the parameters is assessed, while working experimentation aims to obtain new results for studying the computer model.

The transformation of the results of the experiments into information about the object of study includes analysis and evaluation of the results, drawing conclusions and conclusions that provide new information about the object under study. The effectiveness of modeling depends on the creation and maintenance of the necessary documentation for each stage of the process.

In addition, the adequacy of computer models is a key issue. Each model is a means of reflecting objective reality, and the truthfulness of the model is determined by its correspondence to the modeled object. When building a model, simplifications are often made or certain aspects of reality are ignored. Therefore, the conditions and criteria for similarity between the model and the real object must be clearly formulated. In computer models, the conditions for similarity are abstract and the dependence on the physical nature of the processes can make verification of truth more complicated.

Verification of the validity of the computer model is performed at each stage of the process, and in the case of a conceptual model it involves tracing the logical sequence of reasoning and can be performed by specialists who have not participated in the development of the model. When building the model, the correctness of the scheme and the program implementation are checked, and during experimentation, test checks are made with simple examples and calculations. Finally, the final verification of the validity of the model is performed through a careful analysis of the results of the experiments, which should be compared with real situations, when possible [68–70].

3.3. General Information on the Modeling of Power Electronic Devices: Specifics and Features—Unified Approach for Modeling DC/AC Converters Based on a Switching Function

Electrical power converters (including rectifiers, inverters, and DC and AC regulators) have a number of specific characteristics that can be summarized as follows [15,48,65,67,68]:

Power semiconductor elements: The circuits of power converter devices include semiconductor elements that operate in a key mode. This ensures a high efficiency of the converter.

Other components with nonlinear properties: The power circuits of these converters also contain elements such as chokes, transformers, inductors, and electrical machines that have significantly nonlinear characteristics.

Complex control and regulation system: The operation of converter devices is controlled by complex control and regulation systems.

The approach to analyzing, modeling, and designing converters depends on their structure, operating principle, and parameters, taking into account the specific characteristics of each specific circuit. It should be borne in mind that power semiconductor elements usually operate in a key mode in order to achieve high converter efficiency. Depending on the assumptions made when compiling mathematical models, converters are classified into several groups according to the level of complexity:

1. Converters with constant structure and parameters. These converters have a constant structure and parameters during all intervals of the operating period, with only the disturbing effect being subject to discrete changes. These changes are associated with the switching moments of the power semiconductor elements, which are determined by the control system and are known in advance. The processes in such converters are usually described using a system of linear differential equations and piecewise-nonconvergent periodic functions.

2. Converters with changing structure. In these converters, the structure changes during the switching of the power semiconductor elements, with these changes being known in advance and regulated by the control system. Electromagnetic processes can be described in two ways: (a) by systems of linear differential equations whose order and coefficients change at the boundaries between the intervals, or (b) by a single system of linear differential equations with variable coefficients that change at the switching moments.

3. Converters with variable structure depending on switchings. These converters have a structure that changes at previously unknown moments, determined both by the control system and by the voltages and currents in the individual branches of the circuit. The processes can be described in two ways: (a) by systems of linear differential equations whose order and coefficients change at each boundary between the inter-switching intervals, or (b) by a single system of linear differential equations with variable coefficients that change at unexpected switching moments.

4. Converters with variable structures and nonlinear elements. In these converters, the structure changes, and the circuit elements are nonlinear, both active and passive. The electromagnetic processes are described by systems of nonlinear differential equations: (a) for each intercommuting interval with varying order, coefficients, and amount of nonlinearities, or (b) by a single system of nonlinear differential equations with variable coefficients that change at the moments of commutation.

The choice of research method depends on which of these groups the specific electrical energy converter belongs to.

Modeling programs for power electronic devices are classified according to their complexity, capabilities and purpose into two main types [2,3,15,65,70,71]:

Specialized models. These models are based on simplifying assumptions and describe the behavior of a specific type or class of devices in transient and/or steady-state mode. The principle of operation and the switching sequence of power semiconductor elements are known in advance. Specialized models provide more specific and detailed information about a specific type of device.

Universal models. These models cover any operating modes (transient, steady-state, emergency) of electrical energy converters with different structures and parameters. In some cases, power electronics are modeled with universal programs created specifically for power electronic devices, such as PSIM, PLEX and others. In other cases, ready-made software systems are used that are designed for modeling low-power electronics devices, but can also be applied to power electronics.

Modeling DC/AC converters is a key step in the analysis and design of modern power systems. One approach to modeling these devices is the unified model, which uses a switching function to describe the dynamic behavior of the converters. This approach provides high flexibility and accuracy in modeling different types of converters, allowing easy adaptation to different topologies and operating modes.

The main idea of the unified approach is to use a common approach to describe all types of DC/AC converters, including their specific features by appropriately defining the switching function. The switching function is a fundamental element that controls the transition of the system from one operating mode to another. This method offers several main advantages [33,36,55]:

Flexibility and adaptability: Since the approach is based on a common mathematical structure, it can be applied to different types of DC/AC converters (such as inverters, rectifiers and regulators) without the need to create a separate model for each topology. This reduces the complexity of designing new devices and makes the analysis process more efficient.

Easy adaptation to different operating modes: With the help of the switching function, the model can cover different operating modes, including transient, steady and emergency modes. This allows for accurate prediction of the converter behavior under different operating conditions.

Integration with other models and simulation platforms: The unified model can be easily integrated with other software platforms and simulation tools, such as MATLAB/Simulink, allowing for better visualization and optimization of simulation results.

Improved accuracy and control over dynamic processes: The use of a switching function allows for more precise control of the converter dynamics, which is critical for optimizing system efficiency and reliability.

The application of the unified approach to modeling DC/AC converters is particularly useful in the design of devices for renewable energy sources, electric vehicles, and other areas that require high efficiency and reliability in the operation of converters. The unified approach, based on a switching function, offers an efficient way to model DC/AC converters, while providing flexibility and accuracy in analyzing the dynamic behavior of these devices under different operating conditions. This approach is key to improving the design and research processes in the field of power electronics.

3.4. A Specialized Model of Parallel DC/AC Converter in Continuous Current Mode (CCM) and Discontinuous Current Mode (DCM)

By creating this model, the application of the basic methods and principles for building computer models of power electronic devices is shown in a simple way. A parallel thyristor CSI with an active-inductive load according to a full-bridge circuit is considered (Figure 40a). The circuit with the midpoint of the output transformer can be easily reduced to the full-bridge circuit by using numerical coefficients [17]. The example is based on the use of thyristors, but with the assumptions made in the modeling, it is fully relevant when using transistors, taking into account their features.

The assumed positive directions of the state variables (the voltage of the output capacitor C, u; the current of the resonant (input) inductance L_d (L_R), i_d; and the current of the load inductance L, i) are marked on the circuit. The angular frequency of the thyristor control pulses is marked with $\omega = 2\pi f$. It is also known that the circuit in a steady state can operate in one of the following three modes: CCM, BCM and DCM, shown in Figure 43.

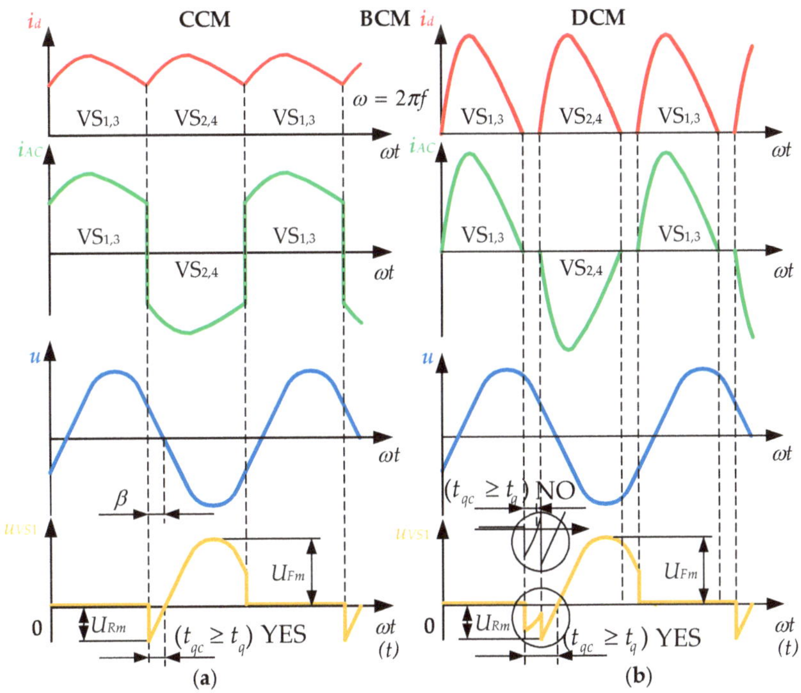

Figure 43. Time diagrams of the main quantities characterizing the different operating modes of a parallel DC/AC converter: (**a**) Continuous Conduction Mode; (**b**) Discontinuous Conduction Mode. Source: Figure by author.

Continuous Conduction Mode (CCM) (Figure 43a). Such a mode occurs at relatively large values of the input choke inductance L_d [15,17]. The input current i_d is continuous and the converter is called a parallel current-source inverter (PCSI). The input current ripples are more pronounced at smaller inductance L_d.

Discontinuous Conduction Mode (DCM) (Figure 43b). Such a mode occurs at relatively small values of the input choke inductance L_d [15]. The input current i_d

is interrupted with pause intervals and the converter is called a parallel resonant inverter (PRI). During the pause, when $i_d = 0$, the load circuit R, L, C is separated from the input power source circuit and independent damping processes occur in it. Two variants of the thyristor voltage development after the conduction interval are possible, depending on the parameters of the inverter circuit. In the favorable case, the thyristor turn-off time scheme t_{qc} consists of the pause interval plus the interval in which the output voltage changes polarity, since the thyristor voltage remains negative during the pause. In the unfavorable case, the latter is not valid, the thyristor turn-off time scheme is less than the pause interval and is too small.

Boundary Conduction Mode (BCM)—the intermediate between CCM and DCM and for the sake of simplicity of consideration can be referred to as DCM.

In the case of an active-inductive load, due to the relatively high quality factor (2–20) of the output oscillatory circuit R, L, C, the output voltage is sinusoidal, although the current acting on the AC circuit (the inverted current i_{AC}) is rich in harmonic components [15,17,33,36].

This study is based on a set of realistic but simplifying assumptions, ensuring both analytical tractability and sufficient proximity to actual operating conditions:

- Active energy consumption occurs solely in the active resistance R of the load circuit.
- Thyristors are modeled as ideal switches, exhibiting no conduction losses or voltage drops in the on-state.
- All other circuit elements are considered lossless, with no parasitic dissipation.
- The power supply is assumed to have zero internal resistance, providing an ideal voltage or current source as required.
- Commutation of current between thyristors is assumed to be instantaneous, without transitional intervals.
- The inductance in the commutation path is neglected, thereby eliminating any influence on switching dynamics.

3.4.1. A Specialized Model of Parallel DC/AC Converter in Continuous Current Mode

When compiling the model, the methods of instantaneous values, switching functions and state variables will be used in combination [15]. In this method, the variables that describe the behavior of each circuit are the voltages of the capacitors and the currents through the inductances: u, i_d, i. From the operation of the circuit, it has been established that in the transient and steady-state operating mode there are three intercommuting intervals. The first is when the thyristors VS_1 and VS_3 conduct, and it is conventionally assumed that this is during the odd half-periods $J = 1; 3; 5; \ldots$ (J is the half-period number). The second is when the thyristors VS_2 and VS_4 conduct during the even half-periods $J = 2; 4; 6; \ldots$ The third is during the pause, when operating in DCM, when none of the thyristors conducts. For each of these intercommuting intervals, it is considered that the moment $t = 0$ coincides with the beginning of the interval. According to the instantaneous value method, the values of the state variables must be adjusted at the switching instants of the power semiconductor elements. This means that the final results for the state variables of the previous intercommuting interval are equal to the initial conditions for the state variables of the next intercommuting interval. The assumed positive directions, in particular for u and i, correspond to the odd half-periods. If the actual directions of the variables are opposite to the assumed positive directions, then in the output results the corresponding quantity will be obtained with a negative sign [15].

The equivalent circuit of the alternating current circuit when the thyristors VS_1 and VS_3 conduct (during the odd half-periods) is shown in Figure 44a. The balance of currents according to Kirchhoff's first law for node A is expressed by the following equation:

$$i_C = C\frac{du}{dt} = i_d - i \tag{81}$$

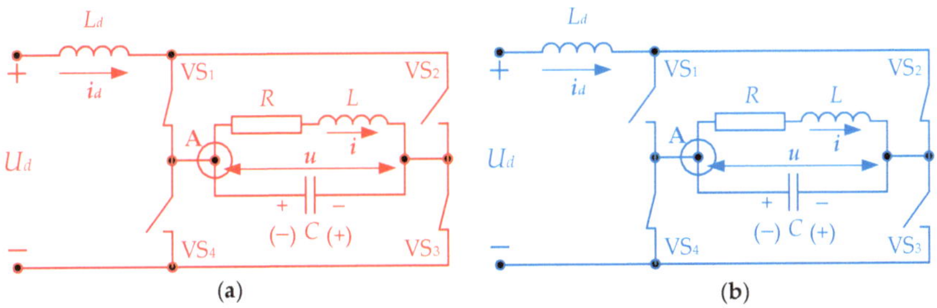

Figure 44. Equivalent circuits of the parallel DC/AC converter when the current in the AC circuit is formed by the joint operation of a pair of semiconductor switches: (**a**) when VS_1 and VS_3 are turned on; (**b**) when VS_2 and VS_4 are turn on. Source: Figure by author.

The voltage balance according to Kirchhoff's second law for the closed loop $+U_d$–L_d–VS_1–u–VS_3–$(-U_d)$ is expressed by the following equation:

$$U_d = L_d\frac{di_d}{dt} + u. \tag{82}$$

The voltage balance according to Kirchhoff's second law for the closed load loop C–R–L is expressed by the equation:

$$u = Ri + L\frac{di}{dt} \tag{83}$$

Equations (81)–(83) are combined into a system of linear differential equations, which is written in normal form (on the left are the derivatives of the variables with respect to time, and on the right are the remaining quantities):

$$\left|\begin{array}{l} \frac{du}{dt} = 0 \cdot u + \frac{1}{C} \cdot i_d - \frac{1}{C} \cdot i + 0 \cdot U_d \\ \frac{di_d}{dt} = -\frac{1}{L_d} \cdot u + 0 \cdot i_d + 0 \cdot i + \frac{1}{L_d} \cdot U_d \\ \frac{di}{dt} = \frac{1}{L} \cdot u + 0 \cdot i_d - \frac{R}{L} \cdot i + 0 \cdot U_d \end{array}\right. , \tag{84}$$

The equivalent circuit of the inverter circuit when the thyristors VS_2 and VS_4 conduct during the even half-cycles is shown in Figure 44b. The balance of currents according to Kirchhoff's first law for node A is expressed by the equation:

$$i_C = C\frac{du}{dt} = -\,i_d - i \tag{85}$$

The voltage balance according to Kirchhoff's second law for the closed loop $+U_d$–L_d–VS_2–u–VS_4–$(-U_d)$ is expressed by the equation:

$$U_d = L_d\frac{di_d}{dt} - u \tag{86}$$

The voltage balance according to Kirchhoff's second law for the closed loop C–R–L, forming the parallel load circuit, is expressed by the Equation (83).

Equations (83), (85) and (86) are combined into a system of linear differential equations written in normal form:

$$\left|\begin{array}{l} \frac{du}{dt} = \quad 0\cdot u - \frac{1}{C}\cdot i_d - \frac{1}{C}\cdot i + \ 0\cdot U_d \\ \frac{di_d}{dt} = +\frac{1}{L_d}\cdot u + \ \cdot i_d + \ 0\cdot i + \frac{1}{L_d}\cdot U_d \\ \frac{di}{dt} = \quad \frac{1}{L}\cdot u + \ 0\cdot i_d - \frac{R}{L}\cdot i + \ 0\cdot U_d \end{array}\right. . \tag{87}$$

To unify the description of the processes in the even and odd half-periods, the switching function shown in Figure 45 is introduced. It is defined as follows:

$$F(\omega\, t) = (-1)^{J+1}, \tag{88}$$

where J is the half-period number. During odd half-periods, when thyristors VS_1, VS_3 are conducting, the value of the switching function is +1, and during even half-periods, when VS_2 and VS_4 are conducting, it is –1.

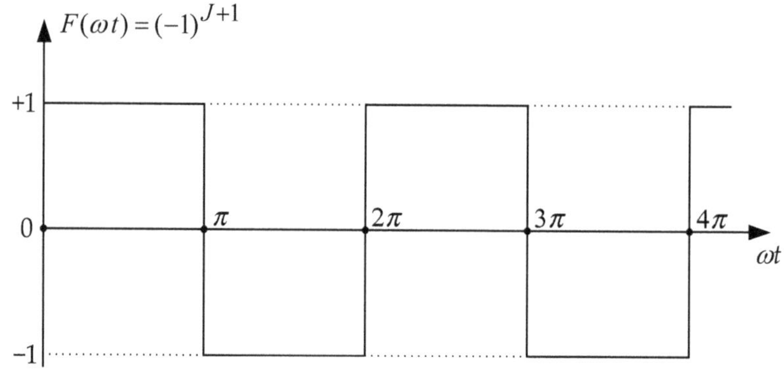

Figure 45. Graphical representation of the switching function used to model a current-source inverter. Source: Figure by author.

Through the switching function, the systems of differential Equations (84) and (87) are summarized into a general system that describes the processes in the parallel DC/AC converter during the conduction of an arbitrary pair of semiconductor switches:

$$\left|\begin{array}{lcl} \frac{du}{dt} &=& 0 \cdot u + \frac{F(\omega\ t)}{C} \cdot i_d - \frac{1}{C} \cdot i + 0 \cdot U_d \\ \frac{di_d}{dt} &=& -\frac{F(\omega\ t)}{L_d} \cdot u + 0 \cdot i_d + 0 \cdot i + \frac{1}{L_d} \cdot U_d \\ \frac{di}{dt} &=& \frac{1}{L} \cdot u + 0 \cdot i_d - \frac{R}{L} \cdot i + 0 \cdot U_d \end{array}\right. . \qquad (89)$$

The system of differential equations in matrix form describing the model of the parallel DC/AC converter in CCM is written as follows:

$$\frac{d[x]}{dt} = [A][x] + [B][e], \qquad (90)$$

where $[x] = [u\ i_d\ i]^T$ is the vector of state variables; $[e] = U_d$ is the vector of the input sources, which in this case, as well as the state matrices of the scheme [A] and [B], are formed by the coefficients of the variables and the input sources in the system (90) and are as follows:

$$[A] = \begin{bmatrix} 0 & \frac{F(\omega\ t)}{C} & -\frac{1}{C} \\ -\frac{F(\omega\ t)}{L_d} & 0 & 0 \\ \frac{1}{L} & 0 & -\frac{R}{L} \end{bmatrix} \text{ and } [B] = \begin{bmatrix} 0 \\ \frac{1}{L_d} \\ 0 \end{bmatrix}. \qquad (91)$$

Thus, the presented mathematical model provides a unified description of the operating modes of the parallel DC/AC converter in CCM, serving as a basis for further analysis and simulation of its dynamic behavior

3.4.2. A Specialized Model of Parallel DC/AC Converter in Discontinuous Current Mode

If the parallel DC/AC converter operates in DCM, then all semiconductor switches are closed during the pause and no current is consumed from the input DC power source, i.e., $i_d = 0$. In this sense, the output (load) circuit R, L, C, which from an electrotechnical point of view is a second-order circuit), remains isolated from the rest of the circuit. There is stored energy in the reactive elements, as a result of the operation of the circuit in the previous intercommuting interval. Thus, free decaying processes develop in the parallel circuit. For the description of the processes, it is assumed that the beginning of the pause coincides with the moment $t = 0$, with the initial conditions of the voltage on the capacitor C being $U(0)_p$, and the current through the inductance $L - I(0)_p$ (Figure 46). Depending on the ratio of the values of the circuit elements, three modes of development of the processes in the circuit are possible: oscillatory, aperiodic, and critically aperiodic [15].

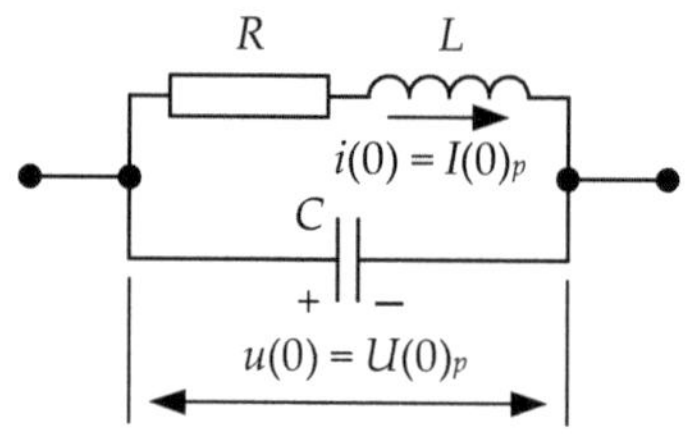

Figure 46. Equivalent replacement circuit of the parallel DC/AC converter during the pause. Source: Figure by author.

1. In the oscillatory mode $\left(R < 2\sqrt{\frac{L}{C}}\right)$, the damping and natural frequency of the circuit are respectively [15],

$$\delta_p = \frac{R}{L} \text{ and } \omega_{0p} = \sqrt{\frac{1}{LC} - \delta_p^2}. \tag{92}$$

The current in the circuit and the voltage across the capacitor are determined by the following expressions:

$$\begin{gathered} i(t) = N_1 e^{-\delta_p t} sin(\omega_{0p} t + N_2), \\ u(t) = L\frac{di(t)}{dt} + Ri(t) = N_1 e^{-\delta_p t}\left(\frac{R}{2} sin(\omega_{0p} t + N_2) + \omega_{0p} L cos(\omega_{0p} t + N_2)\right) \end{gathered} \tag{93}$$

The constants N_1 and N_2 are determined by the initial conditions in the scheme:

$$i(0) = I(0)_p \text{ and } u(0) = U(0)_p, \tag{94}$$

and are defined with:

$$\begin{gathered} N_1 = \frac{U(0)_p - 0.5RI(0)_p}{\omega_{0p} L}\sqrt{1 + \left(\frac{\omega_{0p} LI(0)_p}{U(0)_p - 0.5RI(0)_p}\right)^2}, \\ N_2 = arctg\frac{\omega_{0p} LI(0)_p}{U(0)_p - 0.5RI(0)_p}. \end{gathered} \tag{95}$$

2. In aperiodic mode, the following relationship between the values of the elements of the parallel circuit is valid $R > 2\sqrt{\frac{L}{C}}$, and in this case the roots of the characteristic equation describing the electromagnetic processes in the circuit are:

$$r_{1,2} = -\frac{R}{2L} \pm \sqrt{\frac{R^2}{4L^2} - \frac{1}{LC}}. \tag{96}$$

In this case, the state variables of the circuit under study are as follows:

$$\begin{gathered} i(t) = N_3 e^{r_1 t} + N_4 e^{r_2 t}, \\ u(t) = R(N_3 e^{r_1 t} + N_3 e^{r_2 t}) + L(r_1 N_3 e^{r_1 t} + r_2 N_4 e^{r_2 t}). \end{gathered} \tag{97}$$

The constants N_3 and N_4 are determined by the initial conditions in the scheme and have the following values:

$$N_3 = \frac{U(0)_p - RI(0)_p - r_2 LI(0)_p}{L(r_1 - r_2)} \text{ and } N_4 = \frac{U(0)_p - RI(0)_p - r_1 LI(0)_p}{L(r_2 - r_1)}. \tag{98}$$

3. In the critical aperiodic mode ($R = 2\sqrt{\frac{L}{C}}$), the damping of the circuit is determined by (92), and the current through L and the voltage on C are:

$$\begin{aligned} i(t) &= (N_5 + N_6 t)e^{-\delta_p t}, \\ u(t) &= 0.5R(N_5 + N_6 t)e^{-\delta_p t} + LN_6 e^{-\delta_p t}, \end{aligned} \tag{99}$$

with the constants being defined by:

$$N_5 = I(0)_p \text{ and } N_6 = \frac{U(0)_p - 0.5RI(0)_p}{L}. \tag{100}$$

In CCM and DCM, when conducting a pair of diagonally arranged thyristors, the voltage of a non-conducting thyristor is expressed by:

$$u_{VS} = F(\omega\, t). \tag{101}$$

In the DCM, during the pause interval the voltage of the thyristor, which was conducting immediately before the pause, is:

$$u_{VSon} = \frac{1}{2}(U_d - F(\omega\, t)\cdot u), \tag{102}$$

and the voltage of the thyristor, which was closed immediately before the pause, is:

$$u_{VSoff} = \frac{1}{2}(U_d + F(\omega\, t)\cdot u). \tag{103}$$

The above expressions facilitate the determination of the maximum voltages of the thyristors in the forward (U_{Fm}) and reverse (U_{Rm}) directions in the transient and steady state modes. The schematic turn-off time t_{qc} is determined between the moments of turning off the thyristor and the transition of the voltage on it from negative to positive for the first time. This occurs at discrete moments of time in accordance with the step of numerical integration of the system of differential equations. A more accurate determination of t_{qc} is performed after linear interpolation of the values of u_{VS} for the moment when the thyristor voltage first changed sign, and for the moment one step back.

3.4.3. Integration of the System of Differential Equations Describing the Processes in the Parallel DC/AC Converter

The electromagnetic processes in the DC/AC converter, when conducting a pair of diagonally arranged thyristors, are described by the system of differential Equations (84) or (87), which is of the third order, and during the pause in the

DCM with two variables. An analytical solution of these systems is complex and time-consuming. In this sense, it is more rational to use an appropriate method of numerical integration for solving the system of differential equations. Here is a brief review of the most frequently applied numerical methods used in one of the most widely used mathematical software environments—MATLAB [71,72].

In MATLAB, there are several methods for numerically solving ordinary differential equations (ODEs), which are part of the built-in functions of the software. These methods include both explicit and implicit algorithms for integrating differential equations with different types of conditions and accuracy requirements. MATLAB provides many options for solving systems of ODEs by various numerical methods, which are used depending on the specifics of the problem. Some of the basic methods used in MATLAB are as follows:

1. Euler's Method. Euler's method is a basic, yet fast and simple method for numerically solving ordinary differential equations. It is an explicit method and has been used as a basis for more complex and accurate methods. This method is defined as follows: To solve an ordinary differential equation $\frac{dy}{dt} = f(t,y)$, Euler's method calculates the new value $y(t+h)$ using the expression:

$$y(n+1) = y(n) + hf(t_n, y_n), \tag{104}$$

where h is the time step, and $f(t,y)$ is the function that describes the change in the system.

In MATLAB, this method is implemented using the odeEuler function or implemented manually as part of a custom script.

2. Runge–Kutta Method. The Runge–Kutta method is more accurate than the Euler method and is widely used for numerically solving systems of differential equations. The most popular variant is the fourth-order Runge–Kutta (RK4) method. This method calculates the new value of y by calculating four different estimates of the derivative and combining them. The formula for the fourth-order Runge–Kutta method is:

$$\begin{aligned} k_1 &= hf(t_n, y_n), \\ k_2 &= hf\left(t_n + \tfrac{h}{2}, y_n + \tfrac{k_1}{2}\right), \\ k_3 &= hf\left(t_n + \tfrac{h}{2}, y_n + \tfrac{k_2}{2}\right), \\ k_4 &= hf(t_n + h, y_n + k_3), \\ y_{n+1} &= \tfrac{1}{6}(k_1 + 2k_2 + 2k_3 + k_4). \end{aligned} \tag{105}$$

MATLAB offers the ode45 method, which uses an algorithm based on fourth- and fifth-order Runge–Kutta. It is very accurate and is often used to solve ODEs that require high accuracy, with relatively good computational speed and convergence.

3. Adams–Bashforth Method. The Adams–Bashforth method is an explicit method in a family of multipoint methods that use values from previous steps to calculate the current solution. This makes it very efficient for large numbers of steps. The method uses previously calculated values to predict the value at the next moment, using different weights for the previous steps. In MATLAB, the function ode113 can be used, which implements Adams–Bashforth methods for extended problems and problems with small steps.

Below are the canonical forms of the Adams–Bashforth method: the two–step version (for clarity) and the general k step form. They show how the current approximation is obtained by a weighted linear combination of the derivatives at previous time points.

Two step Adams–Bashforth (AB2):

$$y_{n+1} = y_n + \frac{h}{2}(3f(t_n, y_n) - f(t_{n-1}, y_{n-1})). \tag{106}$$

General k-step Adams–Bashforth:

$$y_{n+1} = y_n + h \sum_{j=0}^{k_{AB}-1} \beta_j f(t_{n-1}, y_{n-1}), \tag{107}$$

where β_j represents the Adams–Bashforth coefficients; y_n is the approximation of the solution $y(t)$ at time t_n; y_{n+1} is the target value at $t_{n+1} = tn + h$; $f(t,y)$—right-hand side of the ODE.

The order the coefficients β_j depend on the chosen order k_{AB} (MATLAB determines them automatically in its variable-order implementation).

Note (for MATLAB): ode113 uses a variable-order PECE scheme (Predict–Evaluate–Correct–Evaluate), with Adams–Bashforth as the predictor and AM (Adams–Moulton) as corrector. This ensures good accuracy with adaptive step-size and tolerance control.

4. Adams–Moulton Method. The Adams–Moulton method is an implicit method that improves accuracy by correction compared to the Adams–Bashforth method. It combines predictions and corrections to solve ODEs. The method uses the same principle as Adams–Bashforth, but with an additional correction of the predicted values to improve accuracy. MATLAB uses the ode15s function to solve problems of high complexity or stiff differential equations.

The method applies the integral form of the ODE and approximates the derivative with a Lagrange polynomial including t_{n+1}.

One-step (Adams–Moulton 1: Backward Euler):

$$y_{n+1} = y_n + hf(t_{n+1}, y_{n+1}), \tag{108}$$

Two-step Adams–Moulton (Adams–Moulton 2: Trapezoidal rule):

$$y_{n+1} = y_n + \frac{h}{2}[f(t_{n+1}, y_{n+1}) + f(t_n, y_n)], \tag{109}$$

General k-step Adams–Moulton:

$$y_{n+1} = y_n + h \sum_{j=1}^{k_{AM}} \alpha_j f(t_{n+1-j}, y_{n+1-j}), \tag{110}$$

where α_j represents the Adams–Moulton weights.

Note (for MATLAB): ode15s automatically adapts the step size and method order, combining Adams–Moulton correction with stability-oriented approaches for stiff systems.

5. Richardson Extrapolation. The Richardson method is a technique used to improve the accuracy of a numerical solution by combining solutions obtained with different integration steps. The Richardson method takes two results with different integration steps and combines the information from them to reduce the error. MATLAB automatically uses similar methods to adaptively control the integration step in functions such as ode45 and others.

If a numerical method produces a solution $S(h)$ for a step h, it can be expressed as:

$$S(h) = S + Ch^p + Oh^{p+1}, \tag{111}$$

where S is the exact solution, C is the constant, p is the order of the method and O is Big O notation.

6. Implicit Methods. Implicit methods are very useful for solving stiff differential equations. They require solving a system of equations at each step and are usually more robust for problems with a high degree of stiffness. An example of implicit methods is the Backward Euler method, which is robust even for very large steps but requires solving nonlinear equations at each step. In Matlab, the ode23s function is used for stiff problems, and is based on implicit methods. A classic example is the Backward Euler method, which is equivalent to the one-step Adams–Moulton method (108).

7. Gauss–Legendre Method;

The Gauss–Legendre method is a numerical integration technique belonging to the class of Gaussian quadrature methods. It achieves very high accuracy by optimally choosing both the integration nodes x_i and the corresponding weights w_i, rather than using equally spaced points as in traditional methods like the trapezoidal or Simpson's rule. The selection of these nodes as the roots of the Legendre polynomial $P_n(x)$ minimizes the integration error for smooth functions. For an integral over the standard interval $[-1, 1]$,

$$\int_{-1}^{1} f(x) \approx \sum_{i=1}^{n} w_i f(x_i), \tag{112}$$

where n is the number of integration points (nodes), x_i represents the Gauss–Legendre nodes, defined as the roots of the Legendre polynomial $Pn(x)$, and w_i are the corresponding weights, computed to ensure that the quadrature formula is exact for all polynomials of degree up to $2n - 1$.

Advantages and Applications: Provides maximum degree of accuracy for a given number of nodes n; particularly effective for smooth functions without singularities in the integration domain; Reduces computational effort compared to equally spaced point methods for the same accuracy; in engineering, it is applied in solving integral equations, evaluating energy functions, and in finite element methods for calculating stiffness matrices and load vectors.

Although MATLAB does not have a built-in Gauss–Legendre quadrature function in its core library, it can be implemented using custom scripts or obtained from specialized toolboxes. For high-precision engineering tasks, Gauss–Legendre integration is preferred when the analytical evaluation of integrals is impractical or impossible.

8. Matrix Exponent Method. The main idea of the matrix exponent method for solving systems of differential equations that model a power electronic device is to use the property of matrices to describe the dynamics of the system. In this context, the system of differential equations is converted into a matrix form, where each state variable is part of a vector, and the interactions between them are represented by a matrix. The method is based on the calculation of the matrix exponent, which allows us to describe the development of the system over time. This provides a solution for how the values of currents and voltages will change under given initial conditions and external influences. Instead of solving differential equations separately for each variable, using the matrix exponent, a general solution is found that unites all variables and their interactions. This is very useful in modeling complex power electronic devices and systems that include multiple components and interactions, such as inverters, converters, etc. The method allows us to quickly analyze the dynamic properties of the device, taking into account both internal parameters (e.g., capacitances and inductances) and external influences that change the state of the system over time.

The matrix exponent method is applied to systems of linear ordinary differential equations in the form:

$$\frac{dx}{dt} = Ax(t),\ x(0) = x_0, \tag{113}$$

The exact solution is given by

$$x(t) = e^{At}x_0, \tag{114}$$

where $x(t)$ is the state vector of the system at time t; A is the system matrix describing the interaction between state variables; e^{At} is the matrix exponential; x_0 is the initial state vector.

This method provides a compact analytical solution for all state variables simultaneously, making it particularly effective for modeling complex power electronic devices such as inverters and converters. In MATLAB, the matrix exponential is computed using the expm function.

MATLAB provides a wide variety of methods for numerically solving ordinary differential equations, which can be selected depending on the specifics of the problem: from simple problems with the Euler method to complex problems with rigid equations solved by implicit methods. For most problems, including non-rigid systems, the most commonly used method is ode45, based on Runge–Kutta. Table 1 presents a comparison of the methods for integrating systems of differential equations used in MATLAB.

Table 1. Comparison of methods for integrating systems of differential equations used in MATLAB.

Method	Description	Advantages	Application
Euler's method	This is a basic and simple method used as a basis for more complex methods.	Easy to implement, fast.	Spatial tasks of low complexity.
Runge–Kuta (RK4)	A more accurate method using the calculation of four different estimates of the derivative.	High accuracy, relatively easy to implement.	Widely used in scientific applications.
Adams–Bashforth	Explicit method using values from previous steps to calculate the current solution.	Effective for tasks with a large number of steps.	Multi-iteration tasks.
Adams–Moulton	An implicit method that improves accuracy by adjusting predictions.	High accuracy and stability.	Tasks requiring high accuracy and stability.
Richardson method	Technique for improving accuracy by combining results with different steps.	Improves accuracy through adaptive pitch adjustment.	Used for adaptive control of the integration step.
Implicit methods	Methods that require solving a system of equations at each step and are stable for hard problems.	Stable for rigid differential equations.	It is used for hard tasks, and is based on implicit methods.
Gauss–Legendre method	Used to integrate differential equations with very high accuracy.	Very accurate.	Tasks with high accuracy requirements.
Matrix Exponent	Method for solving linear differential equations with constant matrices.	Very accurate for linear systems with constant matrices.	Linear dynamic systems, stability analysis.

Source: Table by author.

3.4.4. Implementation of the Model of Parallel DC/AC Converter in Different Programming Environments

In order to study the behavior of parallel DC/AC converters under different operating modes and to optimize both the topology and the control strategies, it is necessary to use simulation models. This section discusses the implementation of a parallel DC/AC converter model in two widely used programming environments—MATLAB/Simulink and LTspice. The choice of these platforms is based on their proven advantages and complementary capabilities [71,73,74]:

- MATLAB/Simulink is an environment with powerful tools for block modeling and simulation of dynamic systems, including real-time control. Here, the emphasis is placed on building a system model of the converter, including control algorithms, feedback loops and the ability to monitor transients within the overall power system.
- LTspice, in turn, is a high-performance electronic circuit simulator that focuses on circuit details such as topology, parameters of electronic components and the shape of the output voltage and current. It allows for precise analyses of the circuit operation at the electronic element level, which is particularly useful in verifying the hardware implementation of the converter.

The simulation performed in both environments allows not only comparison of the results and validation of the model, but also highlights the practical aspects of choosing an appropriate environment depending on the goals of the analysis—be it development of control algorithms or technical design of the power part of the DC/AC converter.

Implementation of a Model of Parallel DC/AC Converter in MATLAB/Simulink

This section discusses the construction of a model of a parallel DC/AC converter in the MATLAB/Simulink environment. The main goal is to create a functional model that allows simulation of the behavior of the converter under different operating modes (discontinuous and continuous current). This implementation provides a basis for further optimization of the parameters, development of control algorithms, and conducting experiments related to tolerance analysis and robustness.

Figure 47 shows a MATLAB/Simulink model of a parallel DC/AC converter that simulates the operation of the device in all possible operating modes. The Simulink model is a direct visual simulation of the system of differential Equation (89), describing the behavior of the parallel inverter. Each row of this system is implemented in a separate circuit, which contains arithmetic blocks and integrators, which implement the determination of the state variables. The switching function $F(\omega t)$ is modeled by a pulse generator, which supplies a control-dependent switching signal between the individual pairs of semiconductor elements, shown in the figure as Control. The output variables u, i_d and i are available for monitoring through the Scope element.

The `Compare To Constant1` ($i_d \geq 0$) block checks whether the input current consumed by the DC power source is greater than or equal to 0, and when the condition $i_d \geq 0$ is met, it forms a unit at its output, otherwise, the state of the block under consideration is 0. This models the condition that corresponds to the physical nature of the electromagnetic processes in the device, namely, that the input current of the inverter cannot be negative. In this way, the continuous and interrupted current modes are combined and it becomes possible to model the parallel DC/AC converter for all possible operating modes.

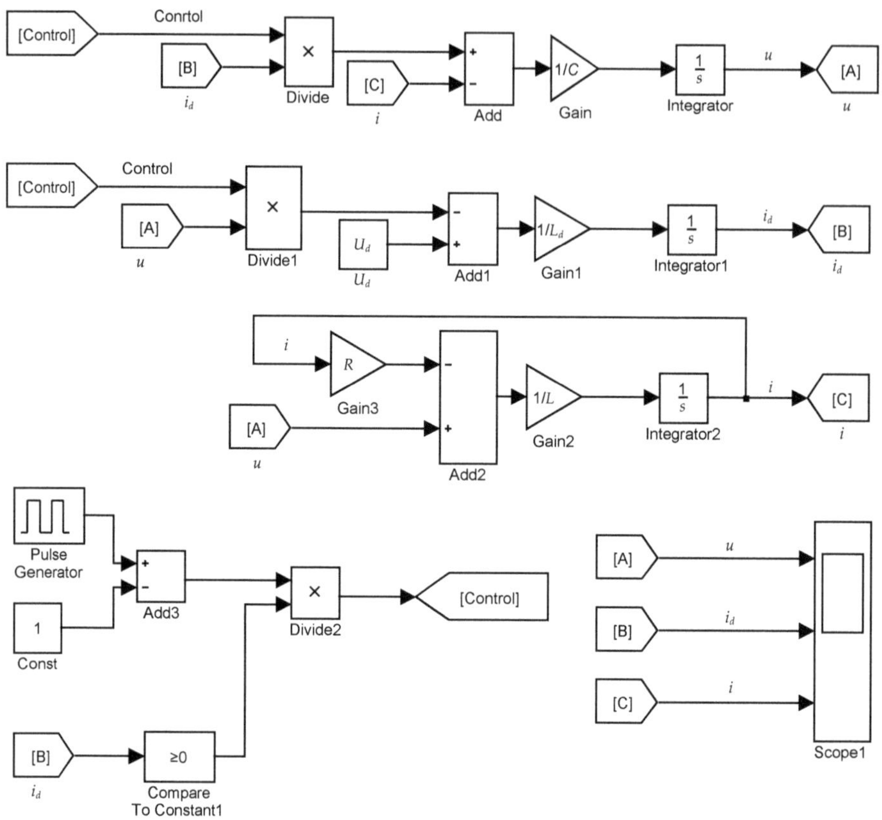

Figure 47. MATLAB/Simulink model of a parallel DC/AC converter. Source: Figure by author.

Implementation of Model of Parallel DC/AC Converter in LTspice

This section discusses the construction and simulation of a parallel DC/AC converter using LTspice, a widely used software platform for simulating electronic circuits. Modeling in LTspice is based on the use of electronic components with their real characteristics and connections between them, and does not require prior formulation of a system of equations, unlike the modeling in the previous section. This makes the approach intuitive and practical, especially when analyzing power electronic converters. Figure 48 presents the simulation model of a parallel DC/AC converter. In the absence of a dedicated thyristor model in the selected simulation environment, each thyristor is substituted by a series connection of a bipolar NPN transistor and a semiconductor diode. This configuration accurately emulates the functional behavior of a real thyristor, reproducing both its conduction and blocking characteristics within the simulated circuit.

The model includes the main elements of a parallel DC/AC converter: semiconductor elements, passive components (inductance, capacitance and resistance), and as control signals implemented through pulse sources. Using LTspice, key electrical quantities such as current and voltage can be monitored, and as the behavior of the system in different operating modes can be analyzed.

The purpose of this simulation is to illustrate the principle of operation of a parallel DC/AC converter and to confirm its functionality in an environment close to the real one, without the need for complex mathematical apparatus. This provides a convenient way to test the idea and subsequent optimization of the circuit parameters.

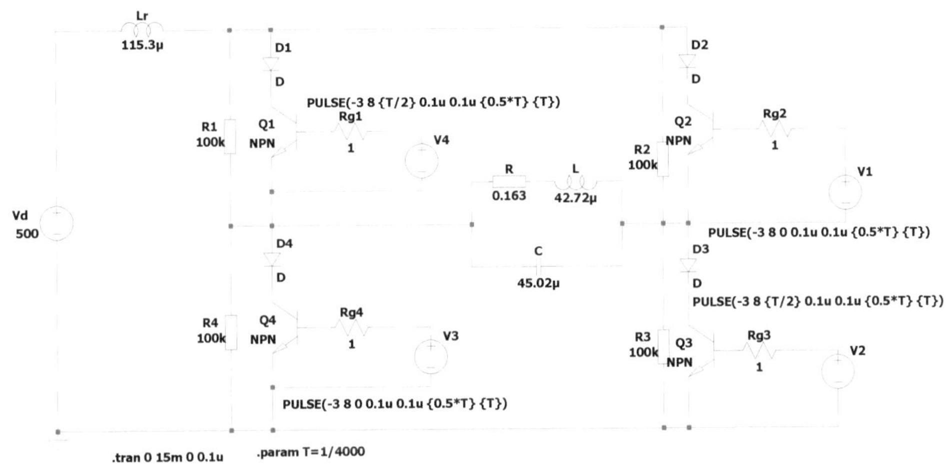

Figure 48. LTspice simulation model of parallel DC/AC converter. Source: Figure by author.

Determining the Integration Step in the Numerical Solution

The integration step is an important parameter in the process of numerically solving systems of differential equations. On the one hand, if the step is too large, the computational process is not accurate and the results differ significantly from reality. If the step is too small, the calculation is slow and more machine time is needed to cover the required time interval for modeling, and in addition, the required disk space for recording the data increases. Therefore, there are optimal values of the integration step, at which the computational process is of good accuracy and is not slow [74].

One of the applied ways to find a suitable integration step is the following. A single calculation is made with one integration step forward in time. Then the same time interval of one step is covered by applying the computational procedure twice, each time with half a step. The results obtained with a single step and with half a step for the same moment in time are compared. If the relative differences for the studied variables are greater than the values previously set by the user for determining the accuracy, it is necessary to return to the initial moment of integration, the calculation to continue with a half step, and the check with a quarter step, etc. The halving of the step continues until the moment of satisfying the set values determining the accuracy. In the event that the relative differences of the results for the studied variables obtained with a single and half step are significantly smaller than the values determining the accuracy, the computational process proceeds slowly with too small a step. Then it is necessary to double the integration step and check the calculation with a single step.

When the eigenvalues of the state matrix [A] differ significantly in order, as is the case in most cases of real circuits of power electronic devices, methods for numerical integration of systems of "hard" differential equations of the "forecast–correction" type are applied. In this case, the derivatives of the studied functions are monitored, which provide information about the speed of the process [15]. At low speed, the calculation is performed with a large step, and at high speed—with a small one. The integration step should be of the order of the smallest time constant, and the duration of the integration interval is related to the largest time constant. The smallest time constants are the circuits for suppressing overvoltages (snubber groups) and the switching circuits of the power semiconductor elements. The time constants of the switching and load circuits are average in value. The largest time constants are the input and filter circuits.

3.5. Specialized Model of Transient and Steady-State Processes in a Series–Parallel DC/AC Converter

The full-bridge circuit of a series–parallel DC/AC converter with an active-inductive load is shown in Figure 49. The assumed positive directions of the state variables (the voltage of the output capacitor C, u; the voltage of the series capacitor C_S, u_S; the current of the input inductance L_d, i_d; and the current of the load inductance L, i) are marked on the circuit. The angular frequency of the thyristor control pulses (the output frequency) is $\omega = 2\pi f$. Similarly to the parallel thyristor DC/AC converter, the circuit can operate in a steady state in three modes: CCM, DCM and BCM.

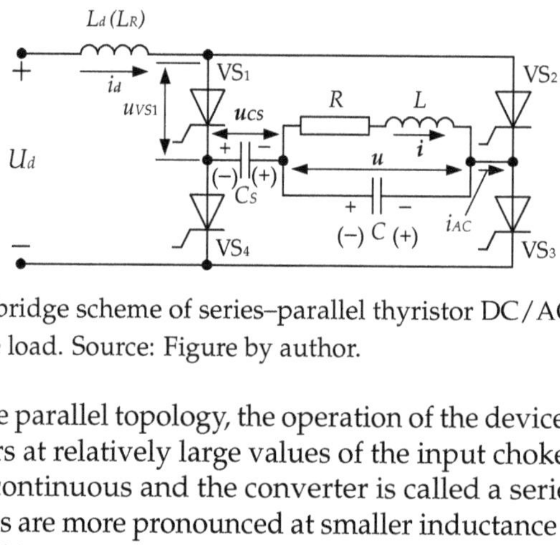

Figure 49. Full-bridge scheme of series–parallel thyristor DC/AC converter with active-inductive load. Source: Figure by author.

Analogous to the parallel topology, the operation of the device in the CCM (shown in Figure 50a) occurs at relatively large values of the input choke inductance L_d. The input current i_d is continuous and the converter is called a series–parallel CSI. The input current ripples are more pronounced at smaller inductance L_d. The operation of the DCM (Figure 50b) occurs at relatively small values of the inductance of the input choke L_d. The input current i_d is interrupted with pause intervals and the converter is called a series–parallel resonant inverter (SPARI). During the pause, when $i_d = 0$, the load circuit R, L, C is separated from the input DC power source circuit and independent damping processes occur in it. During this time, the voltage of the series capacitor C_S–u_S remains almost constant. Two variants of the development of the thyristor voltage after the conduction interval are also possible, depending on the parameters of the DC/AC converter circuit. In the favorable case, the thyristor's circuit turn-off time t_{qc} consists of the pause interval plus the interval in which the inverted voltage (the voltage in the converter bridge diagonal) changes its polarity. In the unfavorable and undesirable case, the thyristor's circuit turn-off time is less than the pause interval and is too small. The BCM occupies an intermediate position between the CCM and the DCM. In the case of the frequently applied active-inductive load, for the same reasons as in the parallel thyristor inverter, the output voltage is practically sinusoidal. The same assumptions are made in the study as in the modeling of the parallel inverter, close to reality.

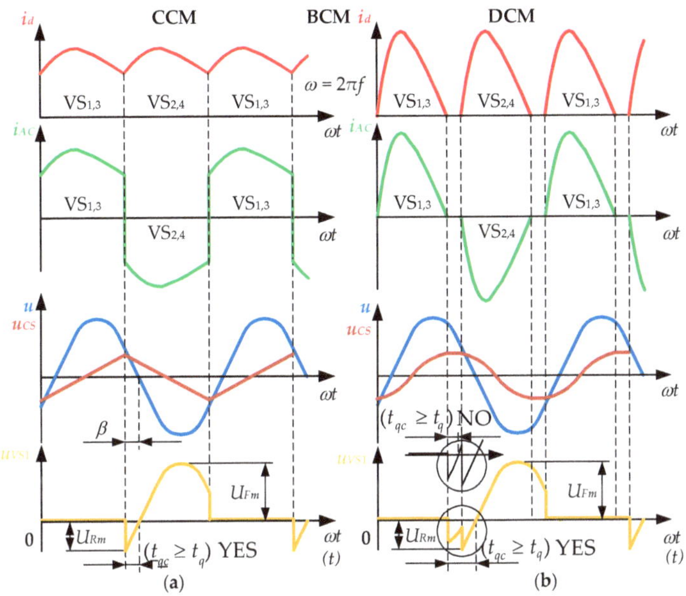

Figure 50. Time diagrams of the main quantities characterizing the different operating modes of the series–parallel DC/AC converter: (**a**) Continuous Conduction Mode; (**b**) Discontinuous Conduction Mode. Source: Figure by author.

The methods of instantaneous values, switching functions and state variables are used in combination. In the transient and steady-state modes, there are three intercommuting intervals: during conduction of the thyristors VS_1 and VS_3, during the odd half-periods, during conduction of the thyristors VS_2 and VS_4 during the even half-periods and during the pause in the DCM, when none of the thyristors conducts. It is considered that the moment $t = 0$ coincides with the beginning of each intercommuting interval.

The equivalent circuits of the inverter circuit, when the thyristors VS_1, VS_3 are conducting during the odd half-cycles, and when VS_2 and VS_4, respectively, are turned on during the even half-cycles, are shown in Figures 51a and 51b, respectively.

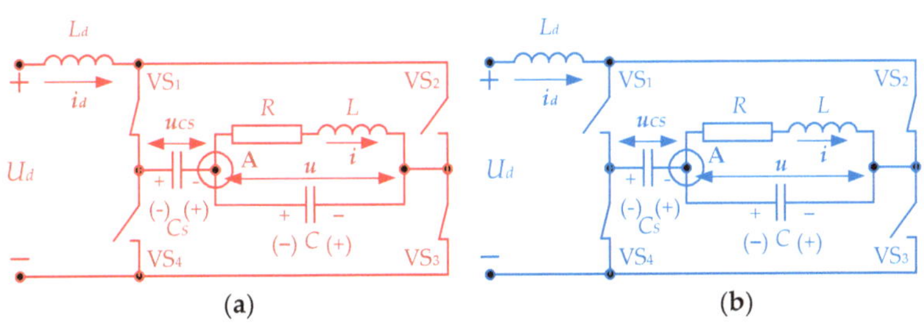

Figure 51. Equivalent circuits of the series–parallel topology in the conduction of a pair of semiconductor switches: (**a**) when VS_1 and VS_3 are turned on; (**b**) when VS_2 and VS_4 are turned on. Source: Figure by author.

During the odd half-periods, when the thyristors VS_1, VS_3 conduct, the system of linear differential equations of the circuit in normal form is expressed by (115). During the even half-periods, when the thyristors VS_2 and VS_4 conduct, the system of linear differential equations of the circuit in normal form is given by (116). In these systems, the first equation follows from the balance of currents according to Kirchhoff's first law for node A. The second equation expresses the relationship between the current and the voltage of the series capacitor C_S. The third equation follows from the balance of voltages according to Kirchhoff's second law for the closed loops formed by the U_d–L_d–C_S–C–thyristors in the on state. The fourth one follows from the balance of voltages according to Kirchhoff's second law for the closed loop C–R–L. Therefore, with the positive directions of the state variables chosen in this way, the following equations are valid for the odd periods:

$$\left|\begin{array}{l}\frac{du}{dt} = 0\cdot u + 0\cdot u_{CS} + \frac{1}{C}\cdot i_d - \frac{1}{C}\cdot i + 0\cdot U_d \\ \frac{du_{CS}}{dt} = 0\cdot u + 0\cdot u_{CS} + \frac{1}{C_S}\cdot i_d + 0\cdot i + 0\cdot U_d \\ \frac{di_d}{dt} = -\frac{1}{L_d}\cdot u - \frac{1}{L_d}u_{CS} + 0\cdot i_d + 0\cdot i + \frac{1}{L_d}\cdot U_d \\ \frac{di}{dt} = \frac{1}{L}\cdot u + 0\cdot u_{CS} + 0\cdot i_d - \frac{R}{L}\cdot i + 0\cdot U_d\end{array}\right. \tag{115}$$

For the even periods of the circuit operation, the corresponding equations are obtained by applying Kirchhoff's laws for the same loops and nodes, taking into account the current directions of the state variables:

$$\left|\begin{array}{l}\frac{du}{dt} = 0\cdot u + 0\cdot u_{CS} - \frac{1}{C}\cdot i_d - \frac{1}{C}\cdot i + 0\cdot U_d \\ \frac{du_{CS}}{dt} = 0\cdot u + 0\cdot u_{CS} - \frac{1}{C_S}\cdot i_d + 0\cdot i + 0\cdot U_d \\ \frac{di_d}{dt} = +\frac{1}{L_d}\cdot u + \frac{1}{L_d}u_{CS} + 0\cdot i_d + 0\cdot i + \frac{1}{L_d}\cdot U_d \\ \frac{di}{dt} = \frac{1}{L}\cdot u + 0\cdot u_{CS} + 0\cdot i_d - \frac{R}{L}\cdot i + 0\cdot U_d\end{array}\right. \tag{116}$$

By analogy with the parallel DC/AC converter, by using the switching function $F(\omega t)$, the systems of differential Equations (115) and (116) are combined into the system (117), which describes the operation of the device during both half-cycles in the CCM.

$$\left|\begin{array}{l}\frac{du}{dt} = 0\cdot u + 0\cdot u_{CS} + \frac{F(\omega t)}{C}\cdot i_d - \frac{1}{C}\cdot i + 0\cdot U_d \\ \frac{du_{CS}}{dt} = 0\cdot u + 0\cdot u_{CS} + \frac{F(\omega t)}{C_S}\cdot i_d + 0\cdot i + 0\cdot U_d \\ \frac{di_d}{dt} = -\frac{F(\omega t)}{L_d}\cdot u - \frac{F(\omega t)}{L_d}u_{CS} + 0\cdot i_d + 0\cdot i + \frac{1}{L_d}\cdot U_d \\ \frac{di}{dt} = \frac{1}{L}\cdot u + 0\cdot u_{CS} + 0\cdot i_d - \frac{R}{L}\cdot i + 0\cdot U_d\end{array}\right. \tag{117}$$

In the DCM during the pause in the output (load) circuit, electromagnetic processes develop, which have already been described in the previous section. At the same time, the current consumed by the DC power source is $i_d = 0$, and the voltage on the series capacitor changes insignificantly, the direction of this change depending on the detuning of the parallel load circuit.

In a similar way, the system of differential Equations (117) is written in matrix form.

3.5.1. Implementation of a Model of a Series–Parallel DC/AC Converter in Matlab/Simulink

The model of the series–parallel inverter is shown in Figure 52. Using block diagrams, Simulink visualizes the system of Equation (117), and also includes a description of the control system and a block for visualization of state variables. The `Compare To Constant1 block` implements modulation of the switching function, thus making it possible to simulate all possible operating modes of the device using a single model. Representing the system of differential equations in this way is convenient compared to using MATLAB code for their solution. When conducting experiments with the model, the same considerations regarding the numerical integration method and the maximum integration step that were discussed in the previous section are valid.

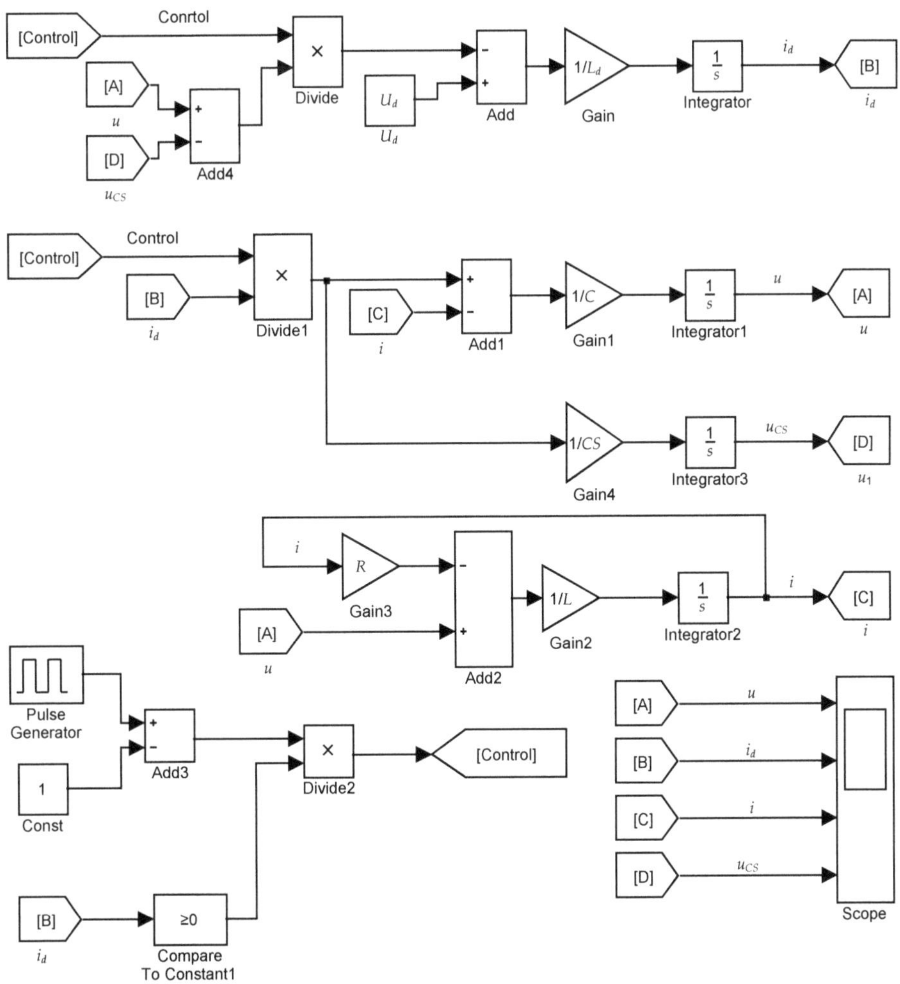

Figure 52. MATLAB/Simulink model of a series–parallel DC/AC converter. Source: Figure by author.

3.5.2. Implementation of a Model of Series–Parallel DC/AC Converter in LTspice

Figure 53 shows the simulation model of a series–parallel DC/AC converter. The scheme combines a series resonant circuit (inductance and capacitor connected in series) with a parallel load resonant circuit to achieve better efficiency and control when working with highly variable loads, such as induction technologies or charging energy storage elements.

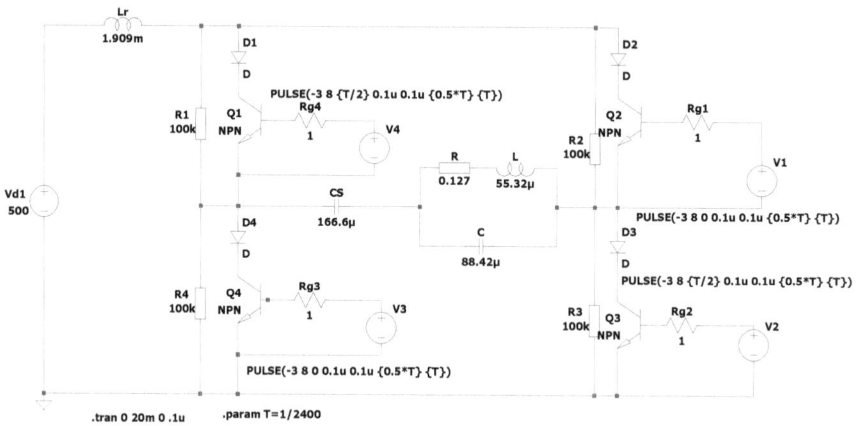

Figure 53. Simulation model of series–parallel DC/AC converter in LTspice. Source: Figure by author.

Control is achieved by pulse signals that turn the transistors on and off at appropriate times to create an alternating voltage of the desired frequency and shape. The diodes in the circuit protect the transistors from reverse voltages, and the resistor dividers improve the convergence of the numerical solution of the model.

3.6. Specialized Model of Transient and Steady-State Processes in a Parallel–Series DC/AC Converter (Inverter with Capacitive Output Voltage Boost)

The full-bridge circuit of a DC/AC converter with capacitive output voltage boost with an active-inductive load is shown in Figure 54. The circuit shows the assumed positive directions of the state variables: the voltage of the output (parallel) capacitor C, u; the voltage of the series capacitor C_L, u_{CL}; the current of the input inductance L_d, i_d; and the current of the load inductance L, i). Similarly to the parallel thyristor inverter, the circuit can operate in a steady state in three modes: CCM, DCM and BCM. The most widely used is CCM, i.e., operation as a CSI.

With an active-inductive load, the output voltage is practically sinusoidal. The study assumes the same, close to reality, assumptions regarding the characteristics of the circuit elements. For the purposes of modeling, the methods of instantaneous values, switching functions and state variables are used in combination. Two intercommuting intervals are considered: with conduction of thyristors VS_1 and VS_3 during odd half-periods and with conduction of thyristors VS_2 and VS_4 during even half-periods. It is assumed that the moment $t = 0$ coincides with the beginning of each intercommuting interval.

The equivalent circuits of the inverter circuit, when conducting thyristors VS_1 and VS_3 during odd half-periods, and VS_2 and VS_4 during even ones, are shown in Figures 55a and 55b, respectively.

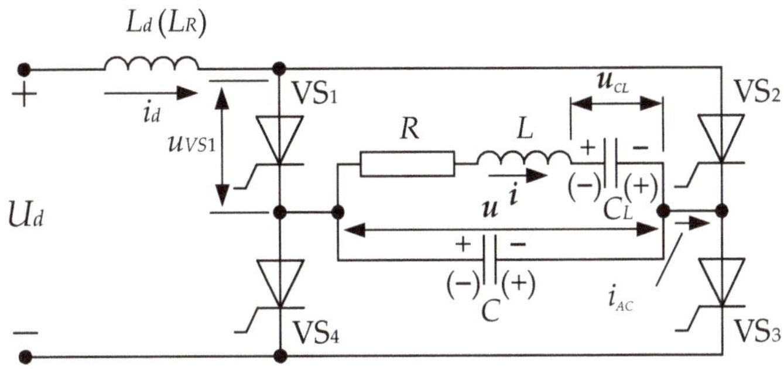

Figure 54. Full-bridge circuit of a DC/AC converter with capacitive increase of the output voltage with an active-inductive load. Source: Figure by author.

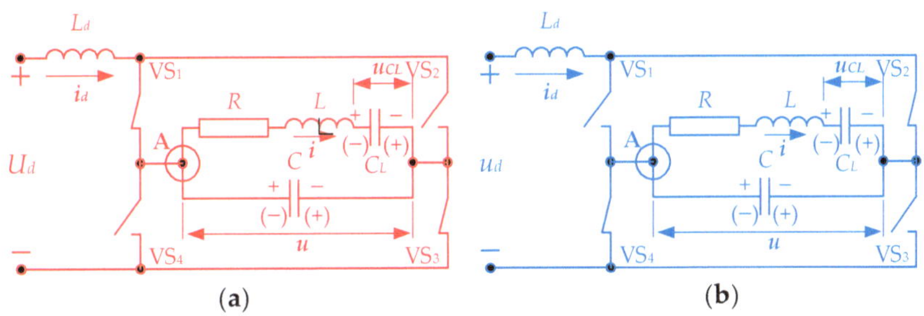

Figure 55. Equivalent circuits of the parallel–series topology in the conduction of a pair of semiconductor switches: (**a**) when VS_1 and VS_3 are turned on; (**b**) when VS_2 and VS_4 are turned on. Source: Figure by author.

In a similar way, by unifying the equations for even and odd half-periods based on the switched function, the generalized system of equations is obtained. In this system, the first equation follows from the balance of currents according to Kirchhoff's first law for node A. The second equation expresses the relationship between the current and the voltage of the series capacitor C_L. The third equation follows from the balance of voltages according to Kirchhoff's second law for the closed loops formed by U_d–L_d–C–thyristors in the turn-on state. The fourth follows from the balance of voltages according to Kirchhoff's second law for the closed loop C–R–L–C_L.

This gives the following system of differential equations that models the transient and steady states in the device:

$$\left| \begin{array}{l} \frac{du}{dt} = \quad 0 \cdot u + \quad 0 \cdot u_{CL} + \frac{F(\omega\, t)}{C} \cdot i_d - \quad \frac{1}{C} \cdot i + \quad 0 \cdot U_d \\ \frac{du_{CL}}{dt} = \quad 0 \cdot u + \quad 0 \cdot u_{CL} + \quad 0 \cdot i_d + \quad \frac{1}{C_L} \cdot i + 0 \cdot U_d \\ \frac{di_d}{dt} = -\frac{F(\omega\, t)}{L_d} \cdot u - \quad 0 \cdot u_{CL} + \quad 0 \cdot i_d + \quad 0 \cdot i + \frac{1}{L_d} \cdot U_d \\ \frac{di}{dt} = \quad \frac{1}{L} \cdot u - \frac{1}{L} \cdot u_{CL} + \quad 0 \cdot i_d - \quad \frac{R}{L} \cdot i + \ 0 \cdot U_d \end{array} \right. \tag{118}$$

In CCM and DCM, when conducting a pair of diagonally arranged thyristors, the voltage of a non-conducting thyristor is expressed by:

$$u_{VS}(t) = F(\omega\, t) \cdot u(t). \tag{119}$$

In the DCM during the pause interval the voltage of the thyristor, which was conducting immediately before the pause, is

$$u_{VSon}(t) = \frac{1}{2}(U_d - F(\omega\, t) \cdot u(t)), \tag{120}$$

and the voltage of the thyristor, which was turn-off immediately before the pause, is:

$$u_{VSoff}(t) = \frac{1}{2}(U_d + F(\omega\, t) \cdot u(t)). \tag{121}$$

The DC supply voltage U_d and the circuit tripping time t_{qc} are determined in the same way as that for the parallel inverter.

3.6.1. Implementation of a Model of Parallel–Series DC/AC Converter in LTspice

Figure 56 shows the simulation model of a parallel–series DC/AC converter-without-reverse-diodes series–parallel inverter. The pulses generated by the PULSE generators (with specified frequency and pulse width parameters) are used to control the transistors Q1, Q2, Q3 and Q4, which serve to change the direction of the current. The diodes (D1, D2, D3, D4) have the role of protecting the transistors from unwanted reverse voltages. This is an important aspect when working with inverters, as it protects the sensitive components of the system from damage. The model in LTspice shows the dynamics of this circuit over time and facilitates tracking its behavior depending on changes in various parameters such as control frequency and circuit parameters.

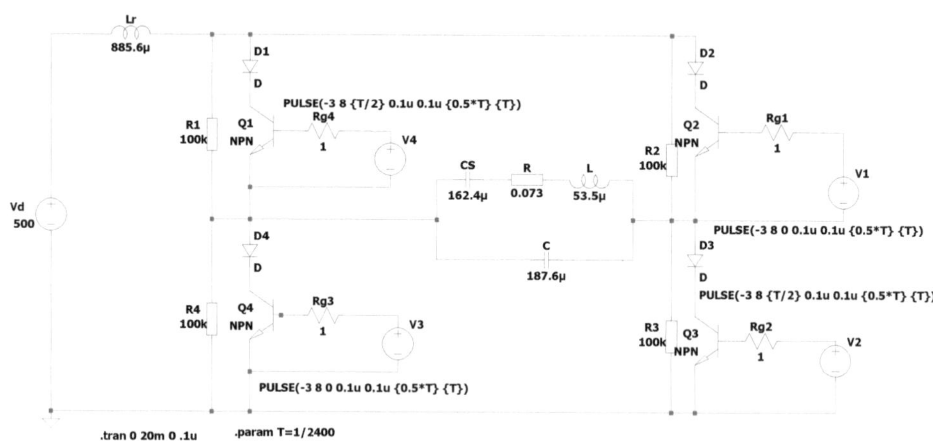

Figure 56. Simulation model of parallel–series DC/AC converter. Source: Figure by author.

3.6.2. Implementation of a Model of Parallel–Series DC/AC Converter in Matlab/Simulink

The parallel–series DC/AC converter model in Matlab/Simulink is shown in Figure 57. The basic principles of building this model are based on the visual representation of the system of differential equations that describe the dynamics of the device. The combination of parallel and series resonant configurations facilitates flexible and efficient control of the conversion of electrical energy.

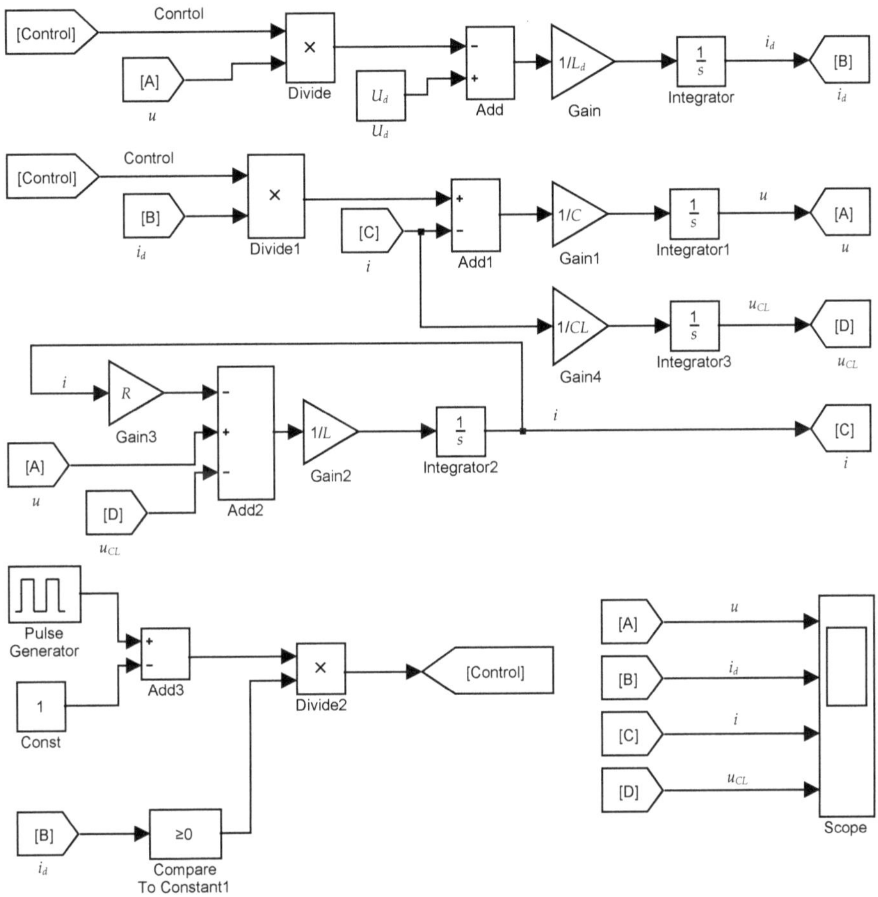

Figure 57. MATLAB/Simulink model of a parallel–series DC/AC converter. Source: Figure by author.

When conducting experiments with the model, the same considerations regarding the numerical integration method and the maximum integration step that were discussed in the previous section are valid.

3.7. Specialized Model of Transient and Steady-State Processes in a Series–Parallel–Parallel–Series DC/AC Converter

The full-bridge circuit of a series–parallel–series DC/AC converter is shown in Figure 58. The ability to adjust the resonance frequency of the series and parallel circuits, through appropriate selection of resonant capacitors, offers significant advantages in achieving higher efficiency and less energy loss in the device. Combining series and parallel resonant elements in one system allows for a wider dynamic range and stability under variable loads. This structure reduces the instability that can occur when the load changes and ensures operability under a wide range of operating conditions.

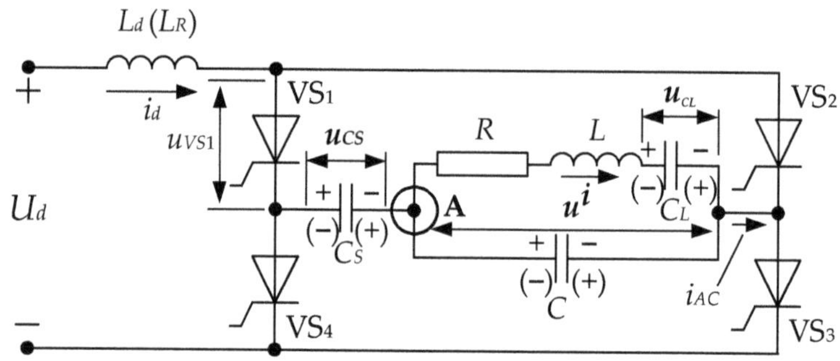

Figure 58. Full-bridge circuit of a series–parallel–parallel–series DC/AC converter. Source: Figure by author.

In a similar way, by combining the equations for even and odd half-cycles based on the switched function, the generalized system of equations is obtained. In this system, the first equation follows from the balance of currents according to Kirchhoff's first law for node A. The second equation expresses the relationship between the current and the voltage of the series capacitor C_L. The third equation follows from the balance of voltages according to Kirchhoff's second law for the closed loops formed by U_d–L_d–C_S–C–thyristors in the on state. The fourth equation follows from the balance of voltages according to Kirchhoff's second law for the closed loop C–R–L–C_L, and the last equation expresses the relationship between the current and the voltage of the series capacitor C_L.

Thus, the following system of differential equations is obtained, which models the transient and steady-state modes in the device:

$$\left|\begin{array}{l} \frac{du_C}{dt} = 0 \cdot u_C + 0 \cdot u_{CL} + 0 \cdot u_{CS} + \frac{F(\omega\, t)}{C} \cdot i_d - \frac{1}{C} \cdot i + 0 \cdot U_d \\ \frac{du_{CL}}{dt} = 0 \cdot u_C + 0 \cdot u_{CL} + 0 \cdot u_{CS} + 0 \cdot i_d + \frac{1}{C_L} \cdot i + 0 \cdot U_d \\ \frac{du_{CS}}{dt} = 0 \cdot u_C + 0 \cdot u_{CL} + 0 \cdot u_{CS} + \frac{F(\omega\, t)}{C_S} \cdot i_d + 0 \cdot i + 0 \cdot U_d \\ \frac{di_d}{dt} = -\frac{F(\omega\, t)}{L_d} \cdot u_C - 0 \cdot u_{CL} - \frac{F(\omega\, t)}{L_d} \cdot u_{CS} + 0 \cdot i_d + 0 \cdot i + \frac{1}{L_d} \cdot U_d \\ \frac{di}{dt} = \frac{1}{L} \cdot u_C - \frac{1}{L} \cdot u_{CL} + 0 \cdot u_{CS} + 0 \cdot i_d - \frac{R}{L} \cdot i + 0 \cdot U_d \end{array}\right. \tag{122}$$

Figure 59 presents a MATLAB/Similink model of a series–parallel–parallel–series DC/AC converter with the ability to operate in CCM and DCM.

The LTspice model of the studied device is not presented, since there are no features there compared to the previous cases, it is simply necessary to repeat the power diagram of the DC/AC converter.

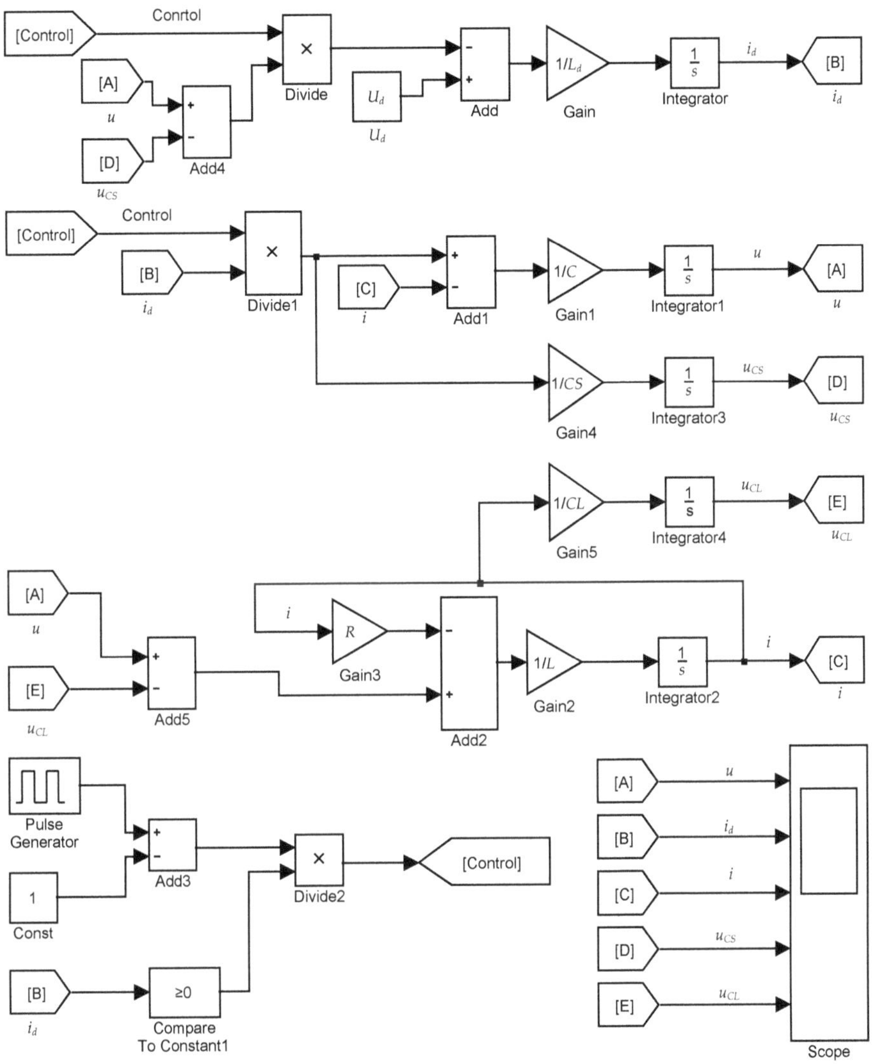

Figure 59. Matlab/Simulink model of a series–parallel–parallel–series DC/AC converter. Source: Figure by author.

3.8. Specialized Model of Series DC/AC Converter

Series resonant inverters (SRIs) are an important part of modern high-frequency electrical energy converters. They are widely used in industries such as induction heating, contactless power supply, medical equipment and telecommunications, where high efficiency, reduced losses and compact dimensions are required [5,20,75]. The basic operating principle of these converters is based on the use of a series LC resonant circuit, which allows operation under natural commutation, minimizing switching losses and electromagnetic interference.

There are two main variants of SRIs: with and without reverse diodes. Variants with reverse diodes provide paths for reverse currents through the switching elements and increase the resistance of the circuit in resonant modes. They improve the safety and reliability of the inverter, especially during dynamic changes in the load. On the other hand, circuits without reverse diodes are characterized by a simpler structure and lower cost, but require careful design of the resonant circuit and control of the operating frequency to avoid unwanted reverse currents and overvoltages [6,19]. On the other hand, RI follower circuits without reverse diodes are mostly thyristor. Due to the widespread use of transistors as semiconductor switches, RI follower circuits are mostly transistor with reverse diodes and for this reason their modeling will be discussed.

The purpose of this section is to consider the modeling of series DC/AC converters. The schematic of a series RIRD is shown in Figure 34. The CCM is considered, using the assumptions introduced earlier.

The switching function is used in the modeling, and the differential equation system, which describes the transient and steady-state modes of operation, is

$$\left| \begin{array}{l} \frac{di_L}{dt} = -\frac{R}{L} \cdot i_L - \frac{1}{L} \cdot u_C + \frac{F(\omega t)}{L} \cdot U_d \\ \frac{du_C}{dt} = \frac{1}{C} \cdot i_L - \ 0 \cdot u_C + \ 0 \cdot U_d \end{array} \right. \tag{123}$$

The implementation of the model in Simulink is given in Figure 60.

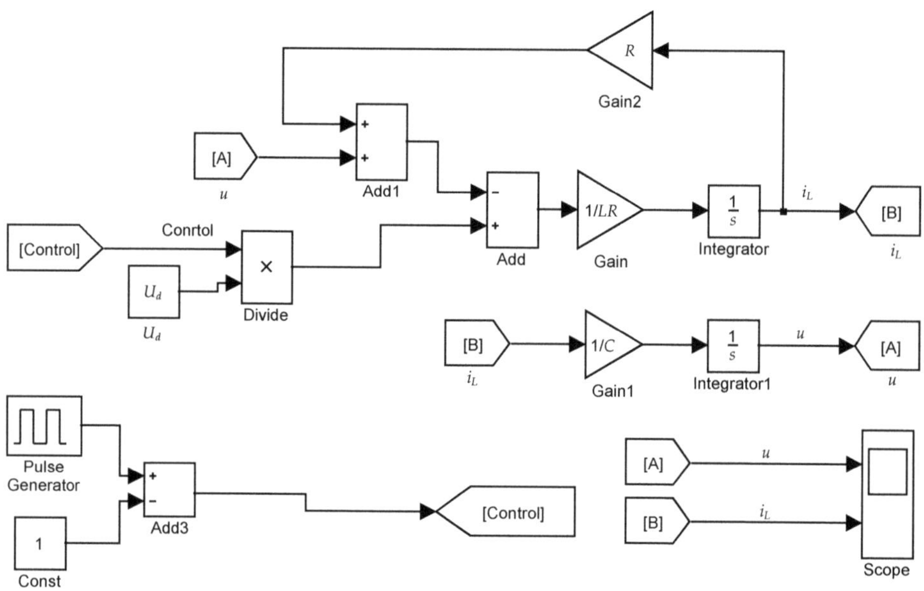

Figure 60. Model of a series transistor RIRD in MATLAB/Simulink. Source: Figure by author.

The implementation of the series RIRD model in Ltspice is shown in Figure 61.

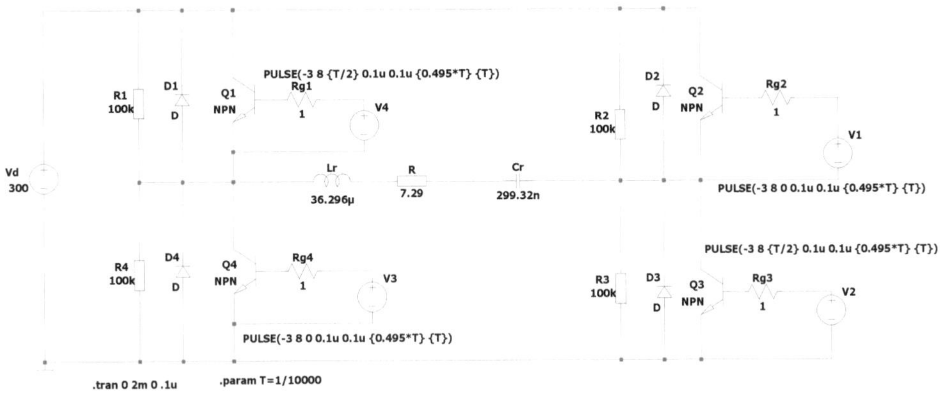

Figure 61. Simulation model of a series resonant DC/AC converter with reverse diodes. Source: Figure by author.

3.9. *Specialized Model of Parallel DC/AC Converters with Reverse Diodes*

Among the various resonant inverter architectures, transistor RIRD and parallel compensated load are of particular importance. Parallel load compensation through various inductive-capacitive resonant circuit topologies facilitates controlling the shape and amplitude of the output current, while reducing current peaks through the semiconductor elements. This contributes to extending the service life of the devices and increasing the overall energy efficiency. The basic circuit of a parallel compensated load RIRD is given in Figure 62.

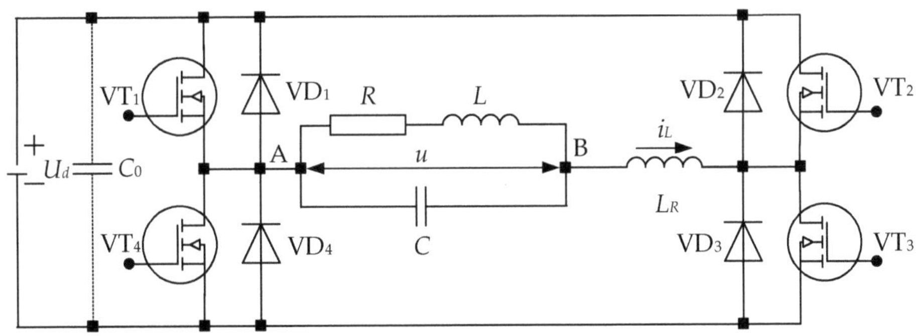

Figure 62. Basic topology of a parallel resonant DC/AC converter with reverse diodes. Source: Figure by author.

The modeling of the circuit is performed analogously to the parallel RI without reverse diodes. A CCM is considered, which is expedient from the point of view of effective use of semiconductor gaps. Through the switching function, the systems of differential equations that describe the processes in the even and odd half-periods of the circuit operation are summarized in a common system that models the parallel RIRD:

$$\left|\begin{array}{llllll} \frac{du}{dt} = & 0 \cdot u + & \frac{1}{C} \cdot i_d - & \frac{1}{C} \cdot i + & 0 \cdot U_d \\ \frac{di_{LR}}{dt} = & -\frac{1}{L_R} \cdot u + & 0 \cdot i_d + & 0 \cdot i + & \frac{F(\omega t)}{L_R} \cdot U_d \\ \frac{di}{dt} = & \frac{1}{L} \cdot u + & 0 \cdot i_d & - \frac{R}{L} \cdot i + & 0 \cdot U_d \end{array}\right. , \tag{124}$$

where u is the voltage across the load and the parallel capacitor C, i_{LR} is the current through the resonant inductance, and i is the current through the load. The parallel RIRD model is implemented in MATLAB/Simulink using the same procedure as in the previous inverters, and its form is given in Figure 63.

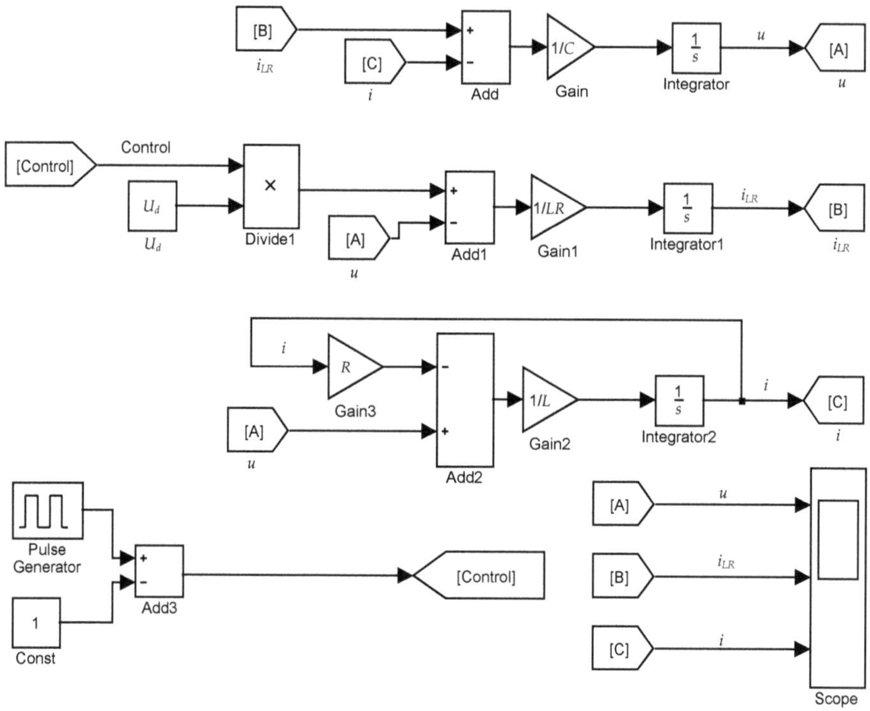

Figure 63. Matlab/Simulink model of parallel resonant DC/AC converter with reverse diodes. Source: Figure by author.

The LTspice model of the parallel RIRD is given in Figure 64.

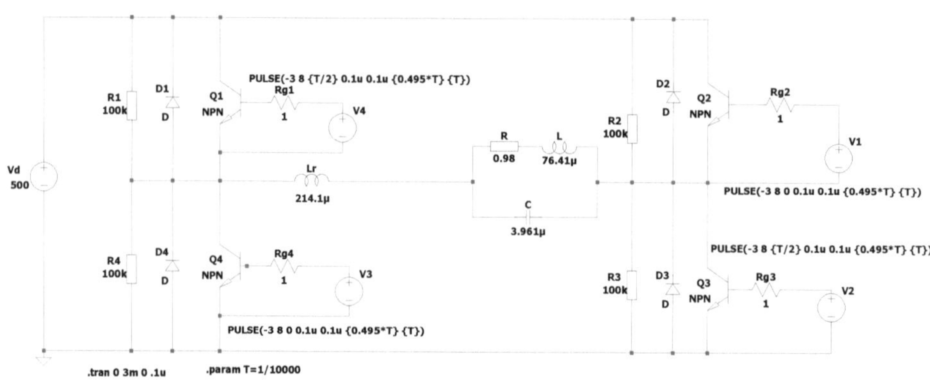

Figure 64. LTspice model of parallel resonant DC/AC converter with reverse diodes. Source: Figure by author.

Another variation of a parallel RIRD is the series–parallel one. To obtain this variant, a series capacitor C_S is added to the basic circuit of a parallel RIRD in

Figure 62, and the type of the inverter AC circuit remains the same as in the circuit in Figure 49. The addition of the series capacitor changes the dynamic response of the circuit, affecting both the voltage and current waveforms. By applying Kirchhoff's laws and the relationships between current and voltage for inductive and capacitive elements, as well as incorporating the switching function $F(\omega t)$, the following system of equations is derived. This system of equations describes the operation of the device during both half-cycles in the CCM, under conditions where the control frequency is either below or above the resonant frequency. This configuration offers advantages in terms of improving system efficiency and load handling, while allowing for finer control over the resonance frequency. Additionally, the series–parallel RIRD circuit is beneficial for applications where higher power handling and more precise regulation of the output voltage are required:

$$\left|\begin{array}{l} \frac{du}{dt} = 0{\cdot}u + 0{\cdot}u_{CS} + \frac{1}{C}{\cdot}i_{LR} - \frac{1}{C}{\cdot}i + 0{\cdot}U_d \\ \frac{du_{CS}}{dt} = 0{\cdot}u + 0{\cdot}u_{CS} + \frac{1}{C_S}{\cdot}i_{LR} + 0{\cdot}i + 0{\cdot}U_d \\ \frac{di_{LR}}{dt} = -\frac{1}{L_R}{\cdot}u - \frac{1}{L_R}u_{CS} + 0{\cdot}i_{LR} + 0{\cdot}i + \frac{F(\omega t)}{L_d}{\cdot}U_d \\ \frac{di}{dt} = \frac{1}{L}{\cdot}u + 0{\cdot}u_{CS} + 0{\cdot}i_{LR} - \frac{R}{L}{\cdot}i + 0{\cdot}U_d \end{array}\right. \tag{125}$$

where u is the voltage across the load and the parallel capacitor C, u_{CS} is the voltage across the series capacitor C_S, i_{LR} is the current through the resonant inductance, and i is the current through the load. The series–parallel RIRD model is implemented in MATLAB/Simulink using the same procedure as in the previous inverters, and its form is given in Figure 65.

The full-bridge circuit of the RIRD with capacitive output voltage boost contains a parallel–series output circuit, such as the one shown in Figure 54. The positive directions of the state variables adopted from the RI circuit were used in the modeling: the voltage of the output (parallel) capacitor C, u; the voltage of the series capacitor C_L, u_{CL}; the current through the resonant (input inductance) L_R, i_{LR}; and the current of the load inductance L, i. Similarly to the parallel thyristor inverter without reverse diodes, the circuit can operate in a steady state in three modes—CCM, DCM and BCM. The most used is CCM, as the control frequency can be below or above the resonant frequency. By using the switching function method and the basic laws of electrical engineering, the system that models the transient and steady modes in the device was obtained:

$$\left|\begin{array}{l} \frac{du}{dt} = 0{\cdot}u + 0{\cdot}u_{CL} + \frac{1}{C}{\cdot}i_{LR} - \frac{1}{C}{\cdot}i + 0{\cdot}U_d \\ \frac{du_{CL}}{dt} = 0{\cdot}u + 0{\cdot}u_{CL} + 0{\cdot}i_{LR} + \frac{1}{C_L}{\cdot}i + 0{\cdot}U_d \\ \frac{di_{LR}}{dt} = -\frac{1}{L_R}{\cdot}u - 0{\cdot}u_{CL} + 0{\cdot}i_{LR} + 0{\cdot}i + \frac{F(\omega t)}{L_R}{\cdot}U_d \\ \frac{di}{dt} = \frac{1}{L}{\cdot}u - \frac{1}{L}{\cdot}u_{CL} + 0{\cdot}i_{LR} - \frac{R}{L}{\cdot}i + 0{\cdot}U_d \end{array}\right. \tag{126}$$

The MATLAB/Simulink implementation of the parallel–serial RIRD model is given in Figure 66.

The full-bridge circuit of a series–parallel–parallel–series RIRD is formed by modifying the basic circuit of a parallel RIRD with the considered configuration of the output circuit. The use of a combination of several resonant circuits in one system facilitates achieving a weaker connection between the load parameters and the equivalent parameters of the replacement series circuit of the AC circuit of the inverter. In this way, operability is guaranteed under a wide range of load changes during a technological process, such as induction melting of metals. As a result of the description of the processes during the different half-periods of the device's operation, the following system of differential equations was obtained, which models the transient and steady-state modes in the device in the CCM:

$$\left|\begin{array}{l} \frac{du_C}{dt} = 0 \cdot u_C + 0 \cdot u_{CL} + 0 \cdot u_{CS} + \frac{1}{C} \cdot i_{LR} - \frac{1}{C} \cdot i + 0 \cdot U_d \\ \frac{du_{CL}}{dt} = 0 \cdot u_C + 0 \cdot u_{CL} + 0 \cdot u_{CS} + 0 \cdot i_{LR} + \frac{1}{C_L} \cdot i + 0 \cdot U_d \\ \frac{du_{CS}}{dt} = 0 \cdot u_C + 0 \cdot u_{CL} + 0 \cdot u_{CS} + \frac{1}{C_S} \cdot i_{LR} + 0 \cdot i + 0 \cdot U_d \\ \frac{di_{LR}}{dt} = -\frac{1}{L_R} \cdot u_C - 0 \cdot u_{CL} - \frac{1}{L_R} \cdot u_{CS} + 0 \cdot i_{LR} + 0 \cdot i + \frac{F(\omega\, t)}{L_R} \cdot U_d \\ \frac{di}{dt} = \frac{1}{L} \cdot u_C - \frac{1}{L} \cdot u_{CL} + 0 \cdot u_{CS} + 0 \cdot i_{LR} - \frac{R}{L} \cdot i + 0 \cdot U_d \end{array}\right. \quad (127)$$

Figure 67 shows the serial–parallel–parallel–serial RIRD model implemented in MATLAB/Simulink.

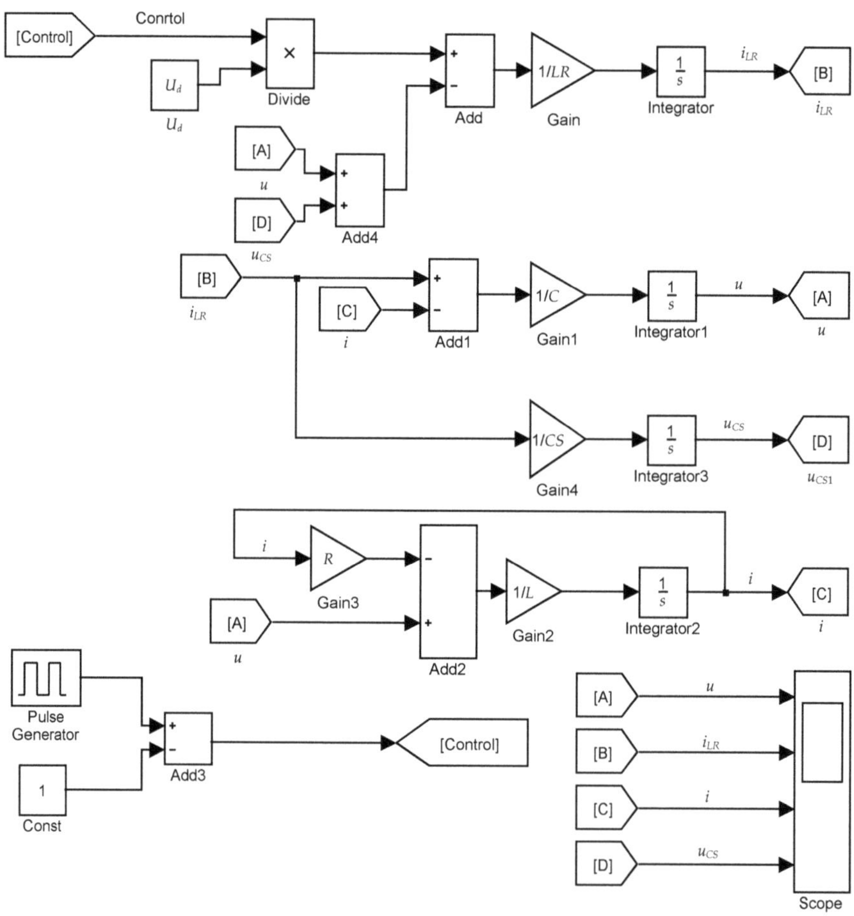

Figure 65. Matlab/Simulink model of a full-bridge series–parallel resonant DC/AC converter with reverse diodes. Source: Figure by author.

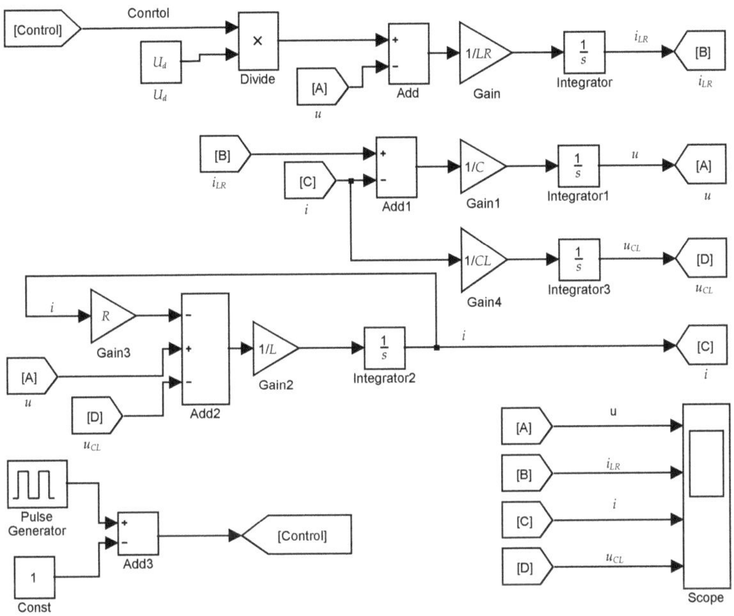

Figure 66. MATLAB/Simulink model of parallel–series resonant DC/AC converter with reverse diodes. Source: Figure by author.

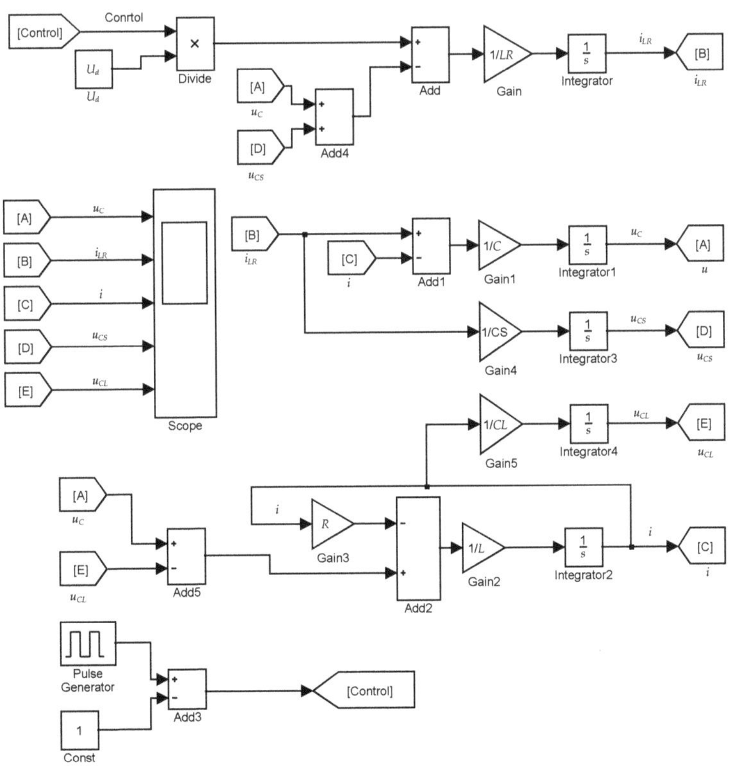

Figure 67. MATLAB/Simulink model of a series–parallel–parallel–serial RIRD. Source: Figure by author.

3.10. Specialized Model of Transient and Steady-State Operation of a Voltage-Source Inverter

Considering the VSI as a special case of an RIRD facilitates analyzing its operating modes from single positions and accordingly modeling these devices using the already created RIRD model [33,55]. The bridge circuit from Figure 17 is considered. Using the switching function, the device model is given by:

$$\left| \frac{di_L}{dt} = -\frac{R}{L}\cdot i_L + \frac{F(\omega t)}{L}\cdot U_d . \right. \tag{128}$$

The Matlab/Simulink model of the VSI is given in Figure 68.

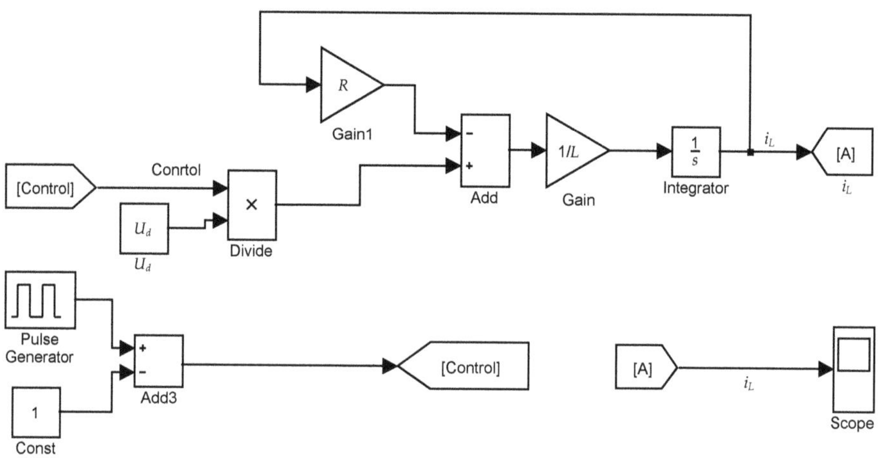

Figure 68. MATLAB/Simulink model of full-bridge VSI. Source: Figure by author.

In a similar way to the previous schematic variations, the model of the VSI in LTspise is built and for this reason it will not be presented.

4. Unified Methodology for the Design of DC/AC Converters

4.1. Introduction

The emergence, rapid development, and diversification of numerous industrial technologies—such as induction heating, melting, alloying, brazing, welding, and other high-frequency processing techniques—have created a sustained demand for advanced power conversion systems capable of precise control and high efficiency. Parallel to these industrial applications, the global trend towards smart grids, sustainable transportation systems, and renewable energy integration has further stimulated research in the field of power electronics, driving innovations in both converter topologies and control strategies.

In recent years, there has also been a notable transfer of knowledge and design experience from heavy industrial environments into consumer-level applications. Contactless power transfer systems, induction cooking appliances, induction boilers and heaters, and high-efficiency lighting systems are prime examples where technological concepts from industrial resonant converters have been successfully adapted for domestic and commercial use [20,76,77]. This trend highlights the versatility of resonant and quasi-resonant topologies and their capacity to meet a broad spectrum of operational requirements.

The availability of a wide range of power semiconductor devices—thyristors, IGBTs, MOSFETs, and their modern wide-bandgap counterparts (SiC, GaN)—together with the diversity of converter configurations, facilitates the implementation of a vast set of control algorithms and operating modes. However, this diversity also presents challenges: each application scenario imposes unique demands on the converter's performance, thermal behavior, efficiency, electromagnetic compatibility, and reliability. This calls for systematic engineering design methodologies that can guide the selection, sizing, and optimization of all key components in a coherent manner.

The literature provides numerous examples of analytical and numerical studies on resonant converters. For example, in [30] static characteristics for converters operating in various modes were obtained through the analysis of idealized series resonant circuits connected to different voltage source configurations. This procedure enables derivation of analytical expressions for the steady-state AC current. Similarly, fundamental frequency analysis has been widely used to characterize steady-state operation for converters within limited frequency ranges [78], producing equations that accurately describe electromagnetic processes and allow direct determination of the circuit elements.

Another significant research direction focuses on model order reduction for resonant converters with a high number of reactive components [79–82]. The high-order equivalent circuits arising in small-signal modeling can be cumbersome to analyze and obscure the physical interpretation of the processes in the power stage. By applying systematic order reduction techniques, simplified yet accurate equivalent models—such as the third-order series resonant converter (SRC) equivalent circuit—have been developed and validated against numerical experiments.

Despite this progress, the literature reveals a distinct gap: there is a lack of comprehensive, methodologically oriented frameworks dedicated to the engineering design of DC/AC converters as a unified class of devices. Most works address specific topologies or operational conditions without providing a generalized methodology that can be adapted across different designs and application domains.

The objective of the present study is therefore to propose a rational, unified engineering design methodology applicable to the three main types of DC/AC converters: current-source inverters (CSIs), voltage source inverters (VSIs), and resonant inverters (RIs). The foundation of this methodology is the unified analytical approach developed by the author, based on a rigorous study of electromagnetic processes in a series RLC circuit [33]. This approach enables consistent analysis across converter types, facilitates direct comparison of performance characteristics, and supports the creation of standardized engineering tools for design, optimization, and control.

4.2. Methodological Basis of the Unified Approach for the Design of DC/AC Converters

In creating unified methodologies for the design of DC/AC converters, first the resonant operating modes are summarized, they use an LC resonant circuit, in which the voltage and current oscillate according to a sinusoidal law. Then the aperiodic modes are considered, because they represent extreme or boundary cases of the inverter operation, in which the system no longer oscillates, but behaves as a damping circuit. This completes the analysis and shows the limits of stable operation and transition to non-resonant states, and thus summarizes all basic schemes of DC/AC converters [15,33,36,55].

Despite the wide variety of induction-based electrical technologies, from the standpoint of classical electrical engineering principles, their electrical representation is most often reduced to a simple series equivalent circuit composed of an active resistance R and a reactive inductive component ωL. This abstraction applies equally well to systems as diverse as induction heating, melting, brazing, and welding equipment, since in all of these cases the load exhibits similar electrical characteristics.

Consequently, the loads seen by the power electronic devices responsible for delivering high-frequency energy to these systems are typically active–inductive in nature, characterized by a low power factor $cos\varphi_L$ [5,16,17,35]. A low power factor implies that a significant portion of the apparent power is associated with reactive energy exchange between the source and the load, rather than being converted into useful work.

From a design perspective, such conditions are undesirable because they increase the current stress on the switching devices and passive components, reduce system efficiency, and demand higher converter ratings for the same level of delivered active power. Therefore, in order to achieve the required active power while minimizing these drawbacks, load compensation becomes essential.

The most common approach is the use of capacitors to counteract the inductive reactance of the load. Depending on how these capacitors are integrated into the load circuit, two principal compensation methods are distinguished:

- Parallel compensation, where the capacitor is connected in parallel with the load, resulting in a series resonant load circuit from the source perspective.
- Series compensation, where the capacitor is connected in series with the load, forming a parallel resonant load circuit with respect to the power source.

The choice between parallel and series compensation is dictated by multiple factors, including the converter topology, desired operating frequency, permissible voltage and current stress levels, and the control strategy to be employed. In either case, the introduction of a compensating capacitor effectively improves the power factor, reduces reactive power flow, and enables more efficient transfer of high-frequency energy from the converter to the load. From the condition for physical implementation of induction heating systems, parallel compensation is mainly used [5,21,83]. Figure 69 shows basic schematics of resonant inverters with electrotechnological applications, using parallel (Figure 69a,b) and series compensation (Figure 69c,d).

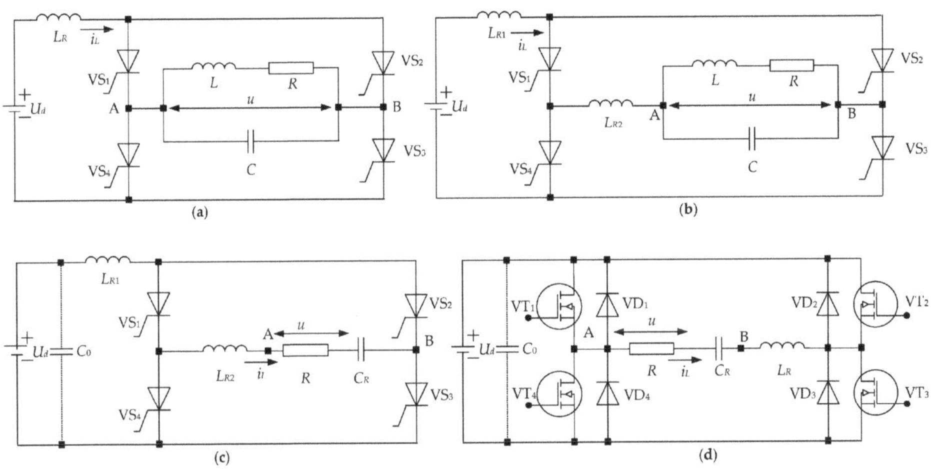

Figure 69. Full-bridge power circuits of resonant DC/AC converters: (**a**) parallel thyristor inverter; (**b**) parallel thyristor inverter with split resonant inductance; (**c**) series thyristor resonant DC/AC converter with split resonant inductance; (**d**) transistor resonant DC/AC converter with reverse diodes. Source: Reprinted from [37], used with permission.

Due to the properties of semiconductor elements, thyristor circuits (Figure 69a–c) are without reverse diodes, while transistor circuits (Figure 69d) are with reverse diodes. In circuits without reverse diodes, it is possible to divide the resonant inductance into two parts (Figure 69b,c), which achieves different levels of voltage load on the thyristors [15,20].

Due to the inherent electrical characteristics of the load, the quality factor of the parallel resonant load circuit typically falls within the range $Q = 3 \div 30$. Such values are common in high-frequency induction systems and indicate a relatively narrow bandwidth of the resonance. As a result, the load circuit performs an important filtering function, effectively attenuating higher-order harmonics in the current of the inverter's AC circuit.

In practical terms, this means that only the fundamental harmonic of the AC circuit current has a significant influence on the load's behavior, while the contribution of all higher harmonics can be neglected for both analysis and design purposes [5,17,19]. This simplification is particularly advantageous because it allows the complex parallel resonant load to be reduced to an equivalent series circuit, which is easier to handle analytically.

In this equivalent representation, the load is modeled by two components connected in series:

- $R_{(1)}$—The active resistance corresponding to the power dissipation in the load when only the fundamental harmonic of the current is considered;
- $X_{(1)}$—The reactive resistance (reactance), representing the energy storage and release associated with the inductive and capacitive elements of the load at the fundamental frequency.

These equivalent parameters are determined using the expressions given in Equation (43). This transformation from a parallel resonant circuit to an equivalent series *R*–*X* model not only simplifies the theoretical treatment but also facilitates the engineering design process, enabling more straightforward calculation of operating conditions, efficiency, and component ratings.

In many practical applications, series compensation is also employed, whereby a compensating capacitor is connected directly in series with the active–inductive load. This configuration is particularly advantageous when modern materials are used that, through the concentration of the electromagnetic field in the inductor, substantially increase the load's power factor $cos\varphi_L$. As a result, the reactive component of the load is reduced, improving the efficiency of energy transfer.

In induction heating and related electrothermal technologies, the loads are typically characterized by low impedance. When series compensation is applied to such loads, a series resonant circuit is naturally formed. This resonance effect enables efficient operation at high frequencies and facilitates better utilization of the inverter's output power.

From a methodological point of view, regardless of whether parallel or series compensation is used, the alternating current circuit of the resonant inverter—no matter how complex in its physical implementation—can be simplified to an equivalent series RLC circuit. This equivalent is valid when considering only the fundamental harmonic of the current at the control frequency, which greatly simplifies analysis and design calculations.

The equivalent circuit corresponding to these considerations is shown in Figure 70. It consists of a resistor R, an inductor L, and a capacitor C connected in series, representing the active, inductive, and capacitive components of the load, respectively. In the special case where the resonant capacitor has an infinitely large value (C→∞), the capacitor is effectively omitted from the circuit. Under such conditions, the equivalent model is also valid for voltage inverters, where operation occurs without a resonant capacitor in the AC circuit.

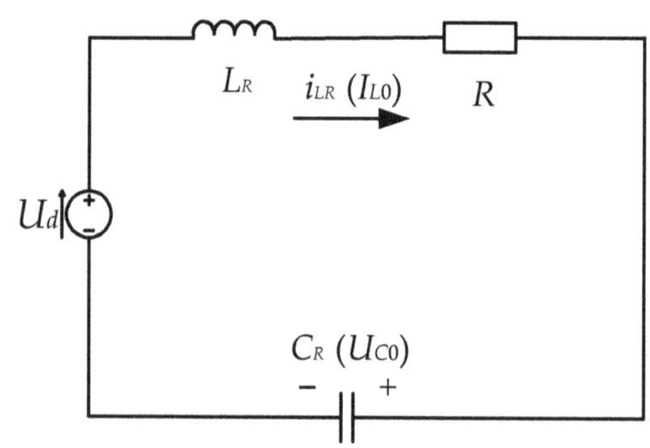

Figure 70. Generalized series equivalent circuit of DC/AC converters. Source: Reprinted from [37], used with permission.

In the various circuit topologies of DC/AC converters, the parameters of the generalized series equivalent circuit from Figure 70 take on specific values depending on the chosen compensation method and the arrangement of the resonant elements:

- Series Compensation: In this configuration, R represents the active component of the load, C_R is the resonant capacitor, and the resonant inductance L_R may be composed entirely of the inductive component of the load L, or it may include an additional inductance L_R connected in series to adjust the resonant conditions. This approach is widely used in induction heating applications where fine-tuning of the resonance is necessary.
- Parallel Compensation: Here, the active resistance $R = R_{(1)}$ is the equivalent active resistance at the fundamental harmonic of the control frequency of the parallel load circuit. The resonant capacitor $C_R = C_{(1)}$, corresponds to the equivalent capacitive reactance at the same harmonic. The resonant inductance L_R may be located entirely in the DC input circuit, entirely in the AC output circuit, or distributed in an arbitrary ratio between them, depending on design requirements.

When complex output matching networks are used to adapt the inverter output to the load parameters, the equivalent circuit of Figure 70 remains valid. However, in such cases the equivalent parameters of the series RLC circuit must be derived using a more sophisticated analytical procedure, which will be presented in the following sections of this chapter.

In certain special configurations of resonant inverters, the resonant function may be performed entirely by a tuned parallel resonant load circuit. A common example of this is when a parallel-compensated load is employed, effectively making the load itself part of the resonant system and influencing the inverter's operating modes.

From the review of the fundamental DC/AC converter topologies used across different applications, it has been established that, for describing the electromagnetic processes in the power circuit, it is most efficient to employ the unified analysis approach presented in Section 2. This methodology allows a consistent mathematical framework to be applied regardless of the specific converter type or compensation method.

When designing this class of power electronic devices, several key coefficients from the unified analysis are utilized:

- Detuning coefficient k—Characterizing the deviation between the control frequency and the natural resonant frequency of the circuit.
- Detuning coefficient of the equivalent resonant circuit ν (also referred to as the aperiodicity coefficient)—Defining the damping characteristics and indicating whether the system is underdamped, critically damped, or overdamped.
- Quasi-frequency coefficient—Providing a normalized measure of frequency that facilitates analytical comparison between different operating regimes.

To further unify the representation of various operating modes, a normalized conduction angle of the semiconductor switches has been introduced: $\lambda = \pi\nu$.

This parameter relates the conduction interval of the switching devices directly to the control frequency, enabling a generalized treatment of both hard- and soft-commutated modes within a single framework.

Another critical characteristic of the equivalent series resonant circuit is the phase angle β between the current in the AC circuit and the output voltage of the DC/AC converter. This angle provides direct insight into the reactive nature of the load and the power factor of the converter system, and is defined as:

$$tg\beta = \frac{X_{CR}}{R}. \tag{129}$$

In resonant DC/AC converters without reverse diodes, the relationship between the control frequency ω and the resonant (or quasi-resonant) frequency ω_0 (Ω_0) of the equivalent AC circuit determines the possible operating modes. According to the unified analysis framework, three distinct modes can be identified, as illustrated in Figure 71 [15,35]:

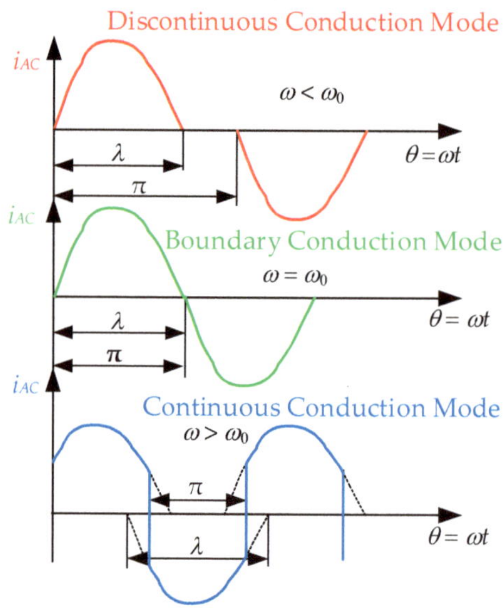

Figure 71. Operating modes of resonant DC/AC converters without reverse diodes. Source: Figure by author.

1. BCM—Boundary Conduction Mode. Occurs when the control frequency exactly matches the natural resonant frequency of the series resonant circuit:

$$\omega = \omega_0 \Rightarrow \lambda = \pi.$$

This mode represents the boundary between continuous and discontinuous current flow in the AC circuit.

2. DCM—Discontinuous Conduction Mode (also referred to as natural commutation). Observed when the control frequency is lower than the resonant frequency:

$$\omega < \omega_0 \Rightarrow \lambda < \pi.$$

In this regime, current through the reactive elements falls to zero before the end of each half-cycle, enabling natural turn-off of the semiconductor devices.

3. CCM—Continuous Conduction Mode (also referred to as forced or hard commutation). Occurs when the control frequency is higher than the resonant frequency:

$$\omega > \omega_0 \Rightarrow \lambda > \pi.$$

In this case, the semiconductor devices are switched while a significant current is still flowing, which requires forced commutation and results in higher switching stress.

From the standpoint of switching behavior in semiconductor devices, these three operating modes can be grouped into two main categories:

- Soft switching—encompassing BCM and DCM, where switching occurs at or near zero current, minimizing losses and electromagnetic interference.
- Hard switching—corresponding to CCM, which in the context of the unified approach is equivalent to the operation of a current-source inverter (CSI).

This classification is of practical importance for converter design, as it directly affects component stress, thermal management requirements, and the achievable efficiency of the system.

RIRD typically operate in the continuous conduction mode (CCM), which is generally more energy-efficient. Depending on the value of the detuning factor ν, two primary operating regimes can be distinguished, namely when the control frequency is below or above the resonant frequency. Figure 72 illustrates the AC circuit current waveforms for both cases. For reference, the waveform in the BCM is also included, providing insight into how the current shape changes with varying ν values.

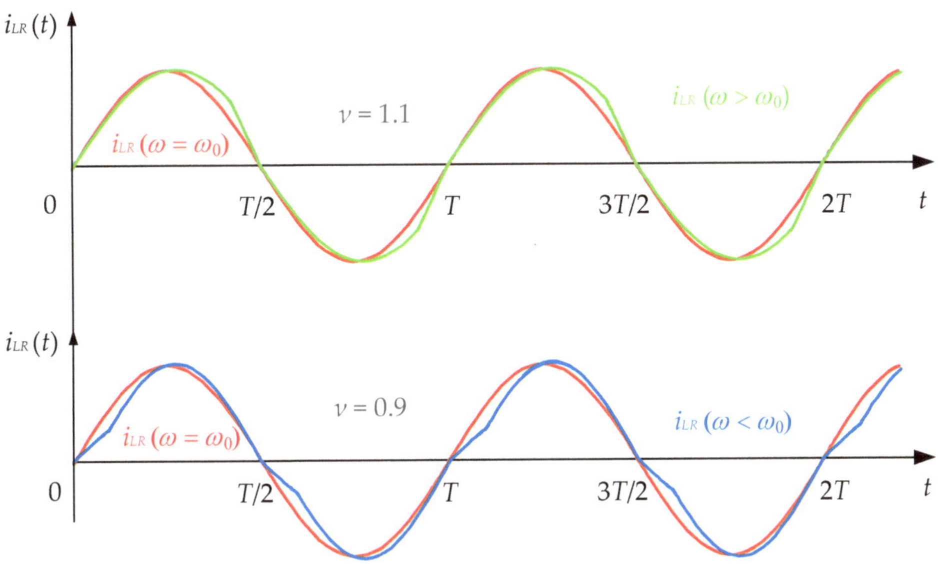

Figure 72. Operating modes of resonant inverters with reverse diodes: operation in BCM ($\omega = \omega_0$) and with control frequency above ($\omega > \omega_0$) and below resonance ($\omega < \omega_0$). Source: Reprinted from [37], used with permission.

For the resonant inverter to function properly—whether it employs fully controllable semiconductor devices (transistors) or partially controllable ones (thyristors)—the equivalent AC circuit (generally of the RLC type) must exhibit a capacitive nature. In practical terms, this means that the current in the circuit leads the inverter output voltage in phase. This requirement can be expressed as follows [61]:

$$tg\beta = \frac{\frac{1}{\omega\ C_R} - (1-\rho)\omega L_R}{R} = \frac{1}{\omega\ C_R R} - \frac{(1-\rho)\ \nu\omega_0}{2\delta}, \tag{130}$$

where the coefficient $\rho = \frac{L_{R1}}{L_{R1}+L_{R2}}$ reflects the distribution of the individual parts of the resonant inductance between the DC and AC circuits of the DC/AC converter.

Using the expression for the resonant frequency of the equivalent series resonant circuit, and performing the necessary mathematical transformations, the following formula for the quantity $\frac{1}{\omega C_R R}$ is derived:

$$\frac{1}{\omega\ C_R R} = \frac{\omega_0^2 + \delta^2}{2\ \nu\delta\omega_0}. \tag{131}$$

In this way, Equation (131) is obtained in the form:

$$tg\beta = \frac{\omega_0^2 + \delta^2}{2\nu\delta\omega_0} - \frac{(1-\rho)\nu^2\omega_0^2}{2\nu\delta\omega_0} = \frac{\omega_0^2\left[1-(1-\rho)\nu^2\right] + \delta^2}{2\nu\delta\omega_0} = \frac{\frac{\omega_0}{\delta}\left[1-(1-\rho)\nu^2\right] + \frac{\delta}{\omega_0}}{2\nu}. \tag{132}$$

Since the unified analysis of DC/AC converters uses the damping factor k, it is more convenient for design purposes to use:

$$tg\beta = \frac{\frac{\pi}{ln\frac{k}{k-1}}\left[1-(1-\rho)\nu^2\right] + \frac{ln\frac{k}{k-1}}{\pi}}{2\nu}. \tag{133}$$

Formula (133) establishes the relationship between the values of the elements of the inverter's AC circuit and the coefficients k and ν, which characterize the operating mode of resonant inverters and form the basis of the unified approach for designing resonant DC/AC converters.

4.3. Transfer Function of Resonant DC/AC Converters

The unified approach to the analysis of DC/AC converters aims to derive the fundamental relationships required for the design of the three principal types of such converters. This method relies on a common representation of the electromagnetic processes occurring in a series RLC circuit, which enables the application of the same analytical framework to current-source, voltage-source, and resonant topologies. By employing this generalized model, it becomes possible to obtain the transfer functions and design equations for different inverter configurations, facilitating direct comparison and parameter optimization across various operating modes.

4.3.1. Transfer Function c of a Resonant DC/AC Converter Operating in Hard Commutation Mode (Current-Source Inverter)

In [15,35] the current-source inverter is treated as a resonant type without reverse diodes, operating under hard commutation conditions (according to Figure 71). Within this framework, permissible ranges for the coefficients k and ν are defined, ensuring that the current waveform in the AC circuit remains close to a rectangular form. The evaluation of waveform proximity is based on two criteria: the phase shift of the fundamental harmonic, $\varphi_{(1)}$, and the current ripple coefficient, K_P. These dependencies, expressed as functions of the parameters k and ν used in the analysis, are illustrated in Figure 73.

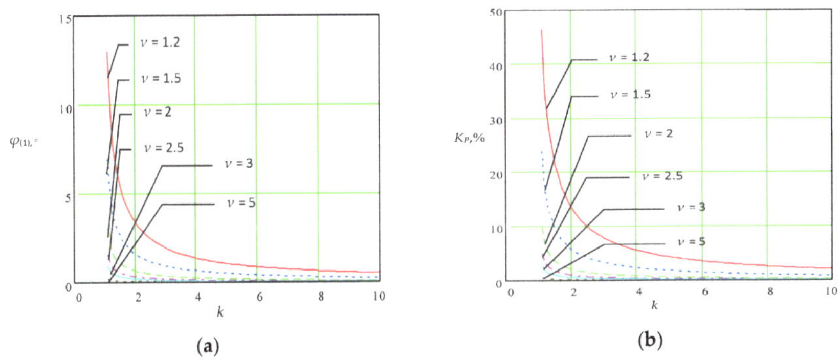

Figure 73. Evaluation of the influence of the equivalent resonant circuit parameters on the shape of the current in the AC circuit of the DC/AC converter: (**a**) phase difference in the first harmonic of the current; (**b**) current ripple coefficient. Source: Reprinted from [37], used with permission.

From the analysis of the graphical dependencies, the following conclusions can be made:

1. Increasing the detuning coefficient k leads the AC circuit current waveform closer to an ideal rectangular profile, for which the characteristic parameters satisfy $\varphi_{(1)} \approx 0$ and $K_P \approx 0$.
2. A higher value of the detuning coefficient ν similarly results in a waveform approaching the ideal rectangular shape, again associated with $\varphi_{(1)} \approx 0$ and $K_P \approx 0$.
3. When k exceeds 2.5 and ν is greater than 3, the deviations become negligible: $\varphi_{(1)}$ falls below 1°, while the ripple coefficient remains under 1%. Under these conditions, the current in the inverter's AC circuit can be considered practically rectangular.

To ensure that a resonant DC/AC converter operates in a hard commutation regime equivalent to that of a CSI, the design should target coefficients within the ranges $k \geq 2.5$ and $\nu \geq 3$. In this case, the RMS output voltage can be determined using the equation derived from the power balance between the input and output of the inverter [15,35]:

$$U_{OUT} = \frac{\pi}{2\sqrt{2}} \frac{U_d}{cos\beta} \approx 1.11 \frac{U_d}{cos\beta}. \quad (134)$$

Here, U_{OUT} denotes the inverter's output voltage, while U_d represents the input voltage supplied by the DC power source. The relationship given in Equation (134) is derived under the assumption of ideal circuit components and the omission of all power losses. In modern DC/AC converters, this approximation is largely valid, as their efficiency typically exceeds 95%. Consequently, the transfer function of the device can be expressed through a relatively simple analytical dependency.

4.3.2. Transfer Function of a Resonant DC/AC Converter Without Reverse Diodes Operating in Soft Commutation Mode (BCM and DCM)

In [35], the resonant inverter without reverse diodes is examined as a particular case of a current-source inverter (CSI). Owing to the specifics of its operating modes, the description of the electromagnetic processes is identical for both the boundary conduction mode (BCM) and the discontinuous conduction mode (DCM), as illustrated in Figure 71. These configurations represent some of the earliest implementations of resonant DC/AC converters and have been the subject of extensive research.

From a design perspective, two parameters are of primary interest: the harmonic content of the AC circuit current and the phase shift of its fundamental harmonic, $\varphi_{(1)}$. As shown in [35,84] for a resonant DC/AC converter without reverse diodes to produce an AC circuit current waveform close to sinusoidal, the coefficients must satisfy $k \geq 1.3$ and $\nu \geq 0.85$. For thyristor-based circuits, operation in DCM is generally recommended, as it facilitates higher switching frequencies.

With the current state of semiconductor technology, resonant DC/AC converters are predominantly implemented using transistors, where operation typically occurs in the boundary conduction mode. In such cases, the control frequency is most often set equal to the resonant frequency of the equivalent series resonant circuit. The RMS value of the output voltage for this type of inverter can be calculated using the following Equation [17]:

$$U_{OUT} = \frac{2\sqrt{2}}{\pi} \frac{U_d}{cos\beta} \approx 0.9 \frac{U_d}{cos\beta} \tag{135}$$

The derived relationship is based on the assumption that, when losses in the circuit components are neglected, the input and output powers of the DC/AC converter are equal.

4.3.3. Transfer Function of a Resonant Inverter with Reverse Diodes Operating in Continuous Conduction Mode

RIRD are most commonly implemented using transistors, as these devices cannot withstand reverse voltage. From an energy-efficiency standpoint, continuous conduction mode (CCM) operation is preferred when the detuning coefficient ν is close to unity, typically in the range 0.85–1.15. For analysis and design purposes, the quasi-boundary method is often applied [85]. This approach determines the transfer function in the boundary conduction mode (BCM), where the circuit behavior is analogous to that of a resonant inverter without reverse diodes, while accounting for the difference between the control frequency and the resonant frequency during AC circuit element selection. As a result, the transfer function coincides with that of resonant inverters without reverse diodes.

The timing diagrams in Figure 72 illustrate the deviation of the actual AC circuit current waveform from that observed in BCM operation ($\nu = 1$). These deviations are found to be minimal and have little impact on design accuracy [33,36].

In summary, despite the diversity of topologies and operating modes, the transfer characteristics of resonant inverters can be reduced to two forms, represented by Equations (135) and (136). A comparison of these transfer functions shows that both are determined by numerical coefficients and the detuning angle of the AC circuit. Notably, in hard-commutated resonant DC/AC converters, the conversion of a dense current pulse—rather than a sinusoidal waveform—results in an output voltage approximately 25% higher than that of soft-commutated resonant DC/AC converters supplied from the same DC source voltage.

4.4. Resonant DC/AC Converters with Complicated Output Circuit

The unified analytical approach to resonant DC/AC converters presented in this work covers operation across different modes and enables the design of such power electronic devices either with a direct load connection or through a matching output transformer.

When it is necessary to enhance the load characteristics of the resonant inverter (RI) and improve its overall energy efficiency, configurations with more complex output circuits are employed. This, in turn, requires a detailed assessment of their influence on the AC circuit, as well as precise optimization of the component parameters in these circuits to achieve the desired performance.

In practical applications, the most widely used complex output configurations are also resonant in nature. The principal types include [15,17]

- Series–parallel output circuit;
- Parallel–series output circuit;
- Series–parallel–parallel–series output circuit.

4.4.1. Series–Parallel Resonant Output Circuit

The schematic diagram of a series–parallel resonant output circuit is presented in Figure 74. To form a complete series–parallel resonant inverter configuration, this circuit is connected to any of the power stages shown in Figure 69 at the points designated A and B. The adopted notation is as follows: R and L represent the parameters of the series equivalent circuit of the load; C_S denotes the series capacitor; and C is the compensation capacitor connected in parallel with the load.

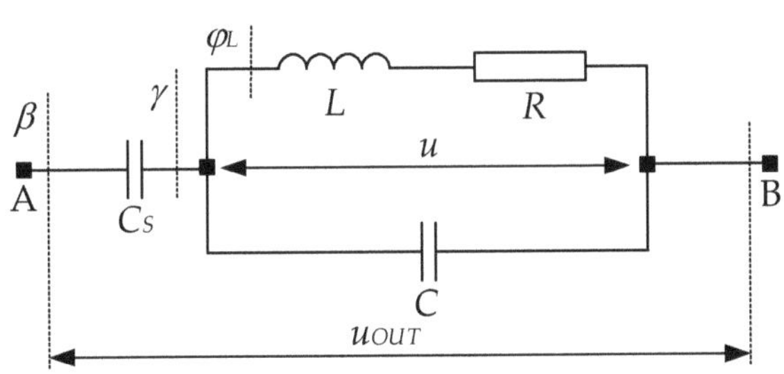

Figure 74. Series–parallel resonant output circuit. Source: Reprinted from [37], used with permission.

When deriving the fundamental relationships for the output circuit under consideration, the ratio between the load voltage U and the inverter output voltage U_{OUT} is introduced, expressed through the coefficient $p = \frac{C_S}{C}$. The parameters used for determining the analytical dependencies are the same coefficients previously defined in the inverter analysis: the detuning k, the detuning ν, and the load power factor $cos\varphi_L$.

To establish the relationship between the load voltage and the inverter output voltage, and to enable the use of any of the already determined transfer characteristics, the circuit shown in Figure 74 must be transformed into an equivalent *R–C* load. Examples of such transformations are presented in [15,33], where the concepts of real and ideal detuning of the respective resonant circuits are applied to express the parameters of the equivalent load. While accurate, this approach considerably complicates both the computational procedures and the physical interpretation of the electromagnetic processes in the power stage.

In this work, an alternative method is proposed, offering substantial simplification in determining analytical relationships and in the design of resonant inverters with complex output circuits. The key tool in this method is the impedance triangle. The graphical constructions illustrating the transformation of the output circuit are shown in Figure 75 [36].

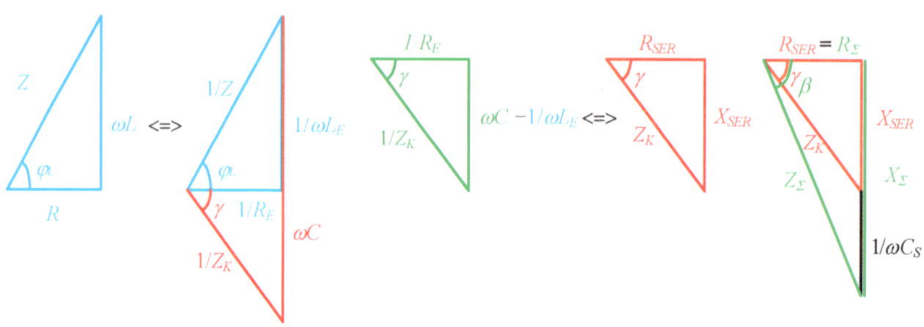

Figure 75. Conversions of a series–parallel resonant output circuit to an equivalent series circuit. Source: Reprinted from [37], used with permission.

The components of the equivalent impedance of the parallel load circuit, comprising R, L, and C, are characterized by the active resistance R_E of the parallel equivalent circuit of the *R–L* load, and by the angle γ between the current in the AC circuit and the load voltage. The sign of γ determines the type of detuning of the load circuit: capacitive or inductive, with $\gamma = 0$ corresponding to resonance. As demonstrated in [17,33] irrespective of the operating mode of the series–parallel resonant inverter (DCM or CCM), capacitive detuning of the load circuit is preferable. Consequently, this case is adopted in the present analysis, and the respective graphical constructions in Figure 75 are prepared for capacitive detuning conditions.

The transformation of the series–parallel circuit is performed in the following sequence. First, the series load circuit (R, L) is converted into an equivalent parallel circuit with the corresponding components R_E and L_E. In this transformation, the angle φ_L between the load current i and the load voltage U is preserved. Due to the compensation capacitor C connected in parallel with the load, the load inductance is offset, resulting in a load resonant circuit equivalent to an active resistance R_E and a capacitive reactance X_E.

The resistances and reactances thus obtained from the parallel equivalent circuit are then transformed into a series equivalent circuit with elements R_{SER} and X_{SER}, while preserving the angle γ during this step. The inclusion of the series capacitor C_S further increases the detuning angle γ of the parallel load circuit. As a result, in the final equivalent circuit, a phase shift angle β is obtained between the current in the AC circuit and the inverter output voltage U_{OUT}.

Through this procedure, the original series–parallel output circuit is ultimately replaced by a series connection of an active resistance R_{Σ} and a capacitive reactance X_{Σ}.

From the balance of active power at the output of the DC/AC converter, the following relationship is obtained [17,36]:

$$\frac{U}{U_{OUT}} = \frac{cos\beta}{cos\gamma}. \tag{136}$$

As a result of the basic relations obtained from the analysis of RI, the angle β is determined by (134). To determine the ratio between the output and load voltages, it is necessary to find an analytical relationship for $cos\gamma$.

From the impedance triangles (Figure 75) we obtain

$$tg\beta = \frac{X_{C_S} + X_{C_{SER}}}{R_{SER}} = \frac{1}{\omega C_S R_{SER}} + tg\gamma. \tag{137}$$

After transformation, Equation (137) takes the form

$$\omega C_S R_{SER} = \frac{1}{tg\beta - tg\gamma}. \tag{138}$$

Similarly, by using the graphical constructions from Figure 75, we obtain

$$tg\gamma = \frac{\frac{1}{X_C} - \frac{1}{X_{L_E}}}{\frac{1}{R_E}} = \omega C R_E - tg\varphi_L. \tag{139}$$

On the other hand, in an equivalent conversion from a series to a parallel replacement circuit, the relationship is valid:

$$R_E = R_{SER}(1 + tg^2\gamma) \tag{140}$$

After substituting (140) into (139) and taking into account (138), the following quadratic equation for the unknown $tg\gamma$ is obtained:

$$(1+p)tg^2\gamma + p(tg\varphi_L - tg\beta)tg\gamma + 1 - ptg\beta tg\varphi_L = 0, \tag{141}$$

where $a = p^2(tg\varphi_L + tg\beta)^2 + 4ptg\beta tg\varphi_L - 4(1+p)$.

The solutions of Equation (141) are as follows:

$$tg\gamma_{1,2} = \frac{p(tg\beta - tg\varphi_L) \pm \sqrt{a}}{2(1+p)}. \tag{142}$$

Of the two solutions of (142), the one that corresponds to the physical nature of the processes in the series–parallel inverter is chosen; i.e., when $C_S \to \infty$, then $cos\beta = cos\gamma$ and, accordingly, $U_{OUT} = U$. In this case, this is the following root of (142):

$$tg\gamma = \frac{p(tg\beta - tg\varphi_L) + \sqrt{a}}{2(1+p)}. \tag{143}$$

Finally, taking into account some trigonometric transformations, we obtain that:

$$\frac{U}{U_{OUT}} = \frac{\sqrt{1+tg^2\gamma}}{\sqrt{1+tg^2\beta}}. \tag{144}$$

Figures 76–79 show the dependence of the ratio between the load voltage and the resonant DC/AC converter output voltage $\frac{U}{U_{OUT}}$, for parameters k, v, p and $cos\varphi_L$, when operating an RI in the considered soft and hard commutation modes. In order to be able to evaluate the transfer characteristic of the series–parallel output circuit, the same values of the coefficient k (with a range of change 1.1–5), the coefficient v (with a range of change from 0.9 to 4), and the load power factor $cos\varphi_L$ are set to four discrete values, respectively, 0.1, 0.2, 0.3 and 0.4, and in all figures the lowest surfaces correspond to the highest value of the load power factor (0.4), and the upper ones to the lowest (0.1).

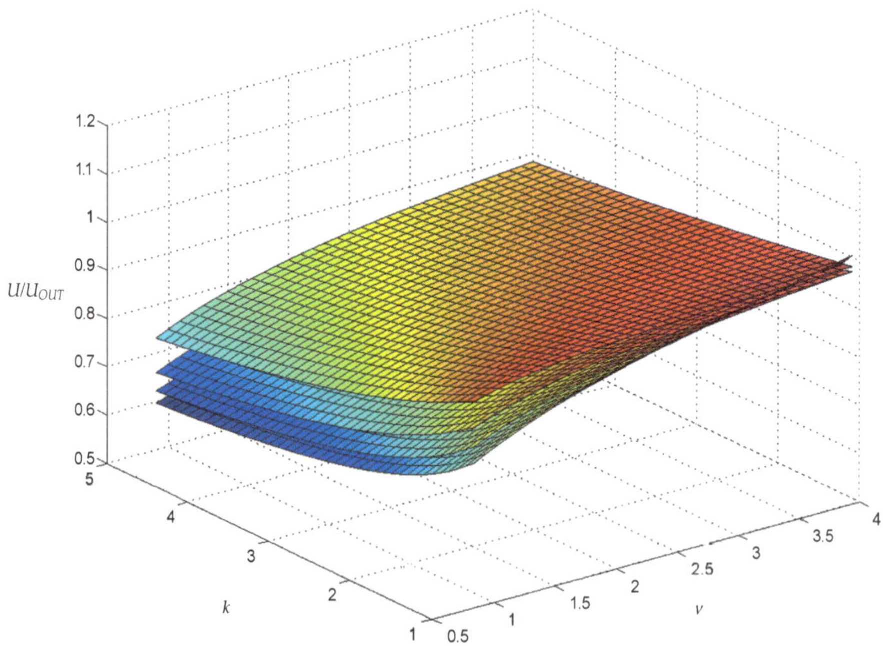

Figure 76. Characteristics of a series–parallel resonant output circuit at parameter value $p = 1$. Source: Figure by author.

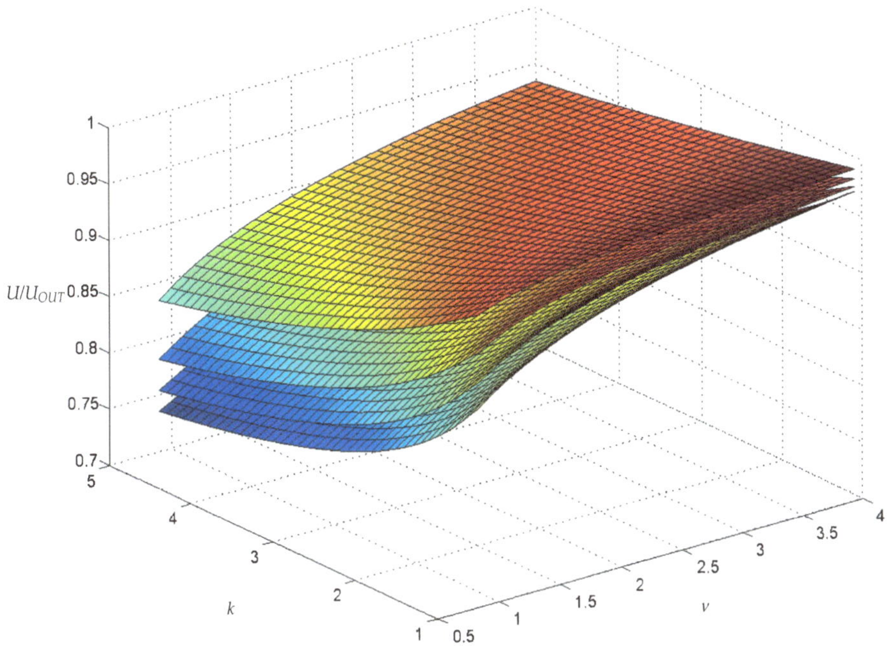

Figure 77. Characteristics of a series–parallel resonant output circuit at parameter value $p = 2$. Source: Figure by author.

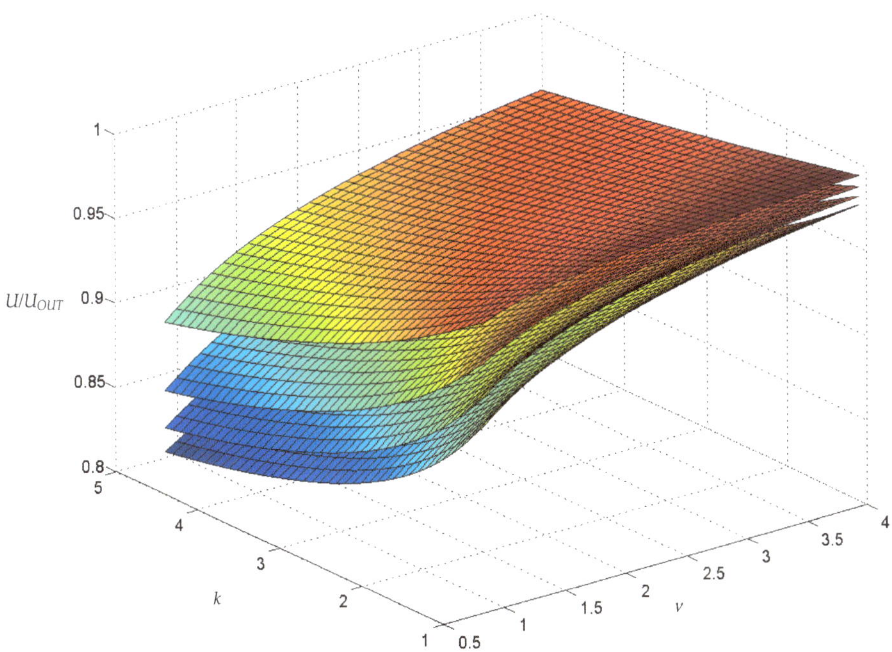

Figure 78. Characteristics of a series–parallel resonant output circuit at parameter value $p = 3$. Source: Figure by author.

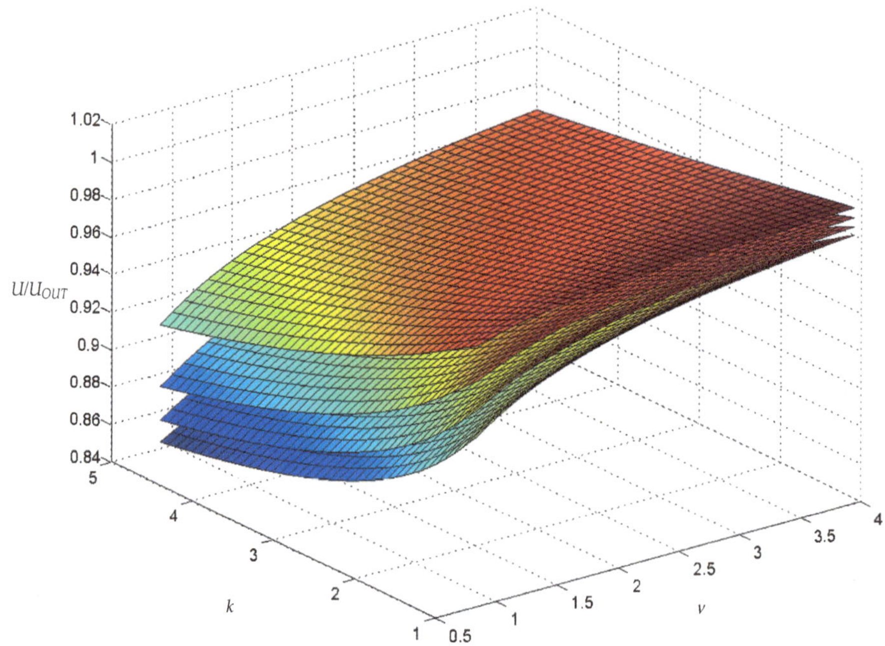

Figure 79. Characteristics of a series–parallel resonant output circuit at parameter value $p = 4$. Source: Figure by author.

The following conclusions were drawn from the analysis of the presented characteristics:

1. With the decrease in $cos\varphi_L$, the ratio U/U_{OUT} increases;
2. The increase in the frequency coefficient ν leads to an increase in the ratio U/U_{OUT};
3. The increase in the damping coefficient k leads to a decrease in the ratio U/U_{OUT};
4. The change in the coefficient k has a stronger influence on the characteristics than the change in the coefficient ν.

From the comparison of the characteristics presented in Figures 76–79, it is established that the increase in the coefficient p leads to an increase in the ratio U/U_{OUT}.

The plotted characteristics confirm the conclusions reported in the literature, namely that the series–parallel resonant output circuit, when operating in the selected mode, is best suited for matching the inverter to a load that requires a voltage lower than the inverter's output voltage.

4.4.2. Parallel–Series Resonant Output Circuit

The circuit diagram of the parallel–series resonant output configuration is shown in Figure 80. To form a parallel–series resonant DC/AC converter, this output stage is connected to any of the power circuits presented in Figure 69 at the points designated A and B. The notation is as follows: R and L represent the parameters of the series equivalent circuit of the load; C_L is the capacitor connected in series with the load; and C is the capacitor connected in parallel with the load.

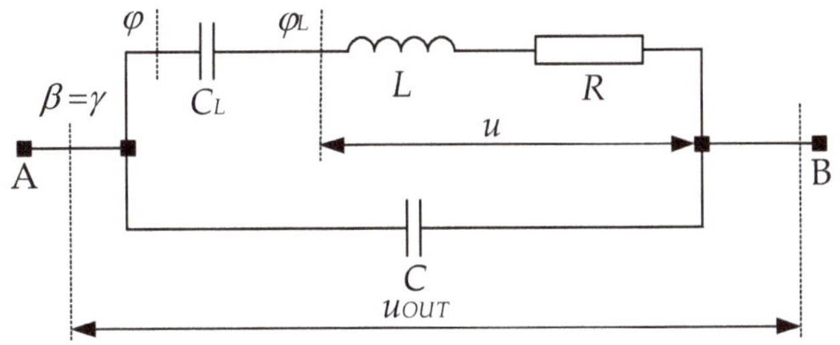

Figure 80. Parallel–series resonant output circuit. Source: Figure by author.

For the purpose of developing engineering design methodologies, it is necessary to derive an analytical equation for the function $\frac{U}{U_{OUT}} = f(s)$, in which the parameters of the functional dependence are the damping factor k, the frequency coefficient ν, and the load power factor $cos\varphi_L$, where $s = \frac{C_L}{C}$.

Similarly to the procedure applied in the previous case, the circuit presented in Figure 80 is transformed step-by-step into an equivalent complex R–C load, making use of the impedance triangle representation of the circuit elements, as illustrated in Figure 81. The series resonant circuit, consisting of R, L, and C_L, can operate under either inductive or capacitive detuning conditions. As noted in [17], inductive detuning is considered more advantageous in terms of reducing the current and voltage stress on the semiconductor switches. Therefore, the present analysis is performed specifically for the case of inductive detuning.

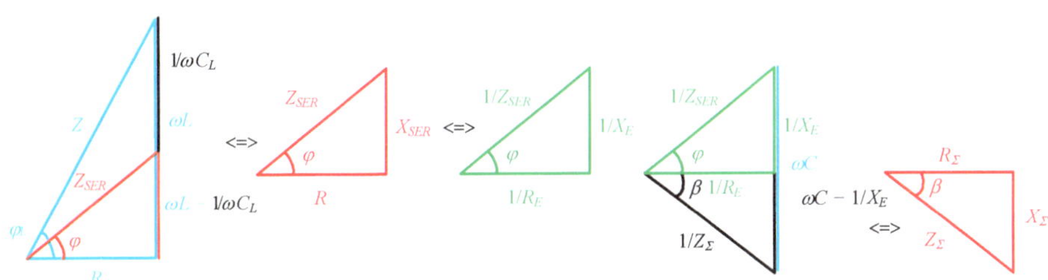

Figure 81. Conversions of a parallel–series resonant output circuit to an equivalent series circuit. Source: Reprinted from [37], used with permission.

In the case of a parallel–series output circuit, the series load components R and L are partially compensated—since the circuit operates under inductive detuning—by the capacitor C_L connected in series with the load. As a result, an equivalent series impedance is obtained, consisting of an active component R and an inductive component $X_{SER} = \omega L - \frac{1}{\omega C_L}$. This impedance is then transformed into an equivalent impedance comprising parallel-connected active $\frac{1}{R_E}$ and $\frac{1}{X_E}$ reactive resistances.

Due to the equivalence of the transformation from a series to a parallel equivalent circuit, the phase angle φ defined between the load current I and the inverter output voltage U_{OUT} is preserved. The resulting active-inductive complex load is subsequently compensated by the parallel capacitor C, producing an equivalent impedance with an active component $\frac{1}{R_E}$ and a capacitive component $\frac{1}{\omega C} - \frac{1}{X_E}$.

The corresponding conductivity triangle is then converted into a resistance triangle, which results in the replacement of the original parallel–series output circuit with a series connection of an active resistance R_{Σ} and a capacitive reactance X_{Σ}. In the final equivalent circuit, the phase shift angle β is established between the current in the AC circuit and the output voltage U_{OUT}.

As demonstrated in [17], the balance of active power in the output circuit leads to the following relationship:

$$\frac{U}{U_{OUT}} = \frac{cos\varphi}{cos\varphi_L}. \quad (145)$$

Since the load power factor $cos\varphi_L$ is known in advance for a given load and is used as a parameter in determining the characteristics of the parallel–series resonant DC/AC converter, the remaining task is to determine the value of $cos\varphi$.

From the constructions of Figure 81, the following equations are obtained:

$$tg\beta = \frac{\frac{1}{X_C} - \frac{1}{X_E}}{\frac{1}{R_E}} = \omega\, CR_E - tg\varphi, \quad (146)$$

from which the following equation is obtained:

$$\omega CR_E = tg\beta + tg\varphi. \quad (147)$$

It is determined similarly that

$$tg\varphi = \frac{X_L - X_{CL}}{R} = tg\varphi_L - \frac{1}{\omega C_L R}. \quad (148)$$

After substituting (147) into (148) and taking into account the relationship between the active resistances between the series and parallel equivalent load circuits $R_E = R(1 + tg^2\varphi)$, after mathematical transformations the following quadratic equation describing $tg\varphi$ was obtained:

$$(1+s)tg^2\varphi - s(tg\varphi_L - tg\beta)tg\varphi + 1 - s\; tg\beta\; tg\varphi_L = 0. \quad (149)$$

The solutions of (149) are

$$tg\varphi_{1,2} = \frac{s(tg\varphi_L - tg\beta) \pm \sqrt{b}}{2(1+s)}, \quad (150)$$

where $b = s^2(tg\varphi_L + tg\beta)^2 + 4s\; tg\beta\; tg\varphi_L - 4(1+s)$.

From the two solutions of Equation (150), the one that corresponds to the physical nature of the processes in the parallel–series RI is chosen; i.e., when $C_L \to \infty$, then $cos\varphi = cos\varphi_L$ and $U_{OUT} = U$. The following root of (150) corresponds to this condition:

$$tg\varphi = \frac{s(tg\varphi_L - tg\beta) + \sqrt{a}}{2(1+s)}. \quad (151)$$

Using trigonometric transformations, we get that

$$\frac{U}{U_{OUT}} = \frac{\sqrt{1+tg^2\varphi_L}}{\sqrt{1+tg^2\varphi}}. \tag{152}$$

Figures 82–85 show the dependence of $\frac{U}{U_{OUT}}$, for parameters k, ν, s and $cos\varphi_L$, when operating an RI in soft and hard commutation modes. In order to be able to evaluate the transfer characteristic of the parallel–series output circuit, the same values of the coefficient k (with a range of change 1.1–5), the coefficient ν (with a range of change from 0.9 to 4), and the load power factor $cos\varphi_L$ are set to 4 discrete values, respectively, 0.1, 0.2, 0.3 and 0.4, and in all figures the lowest surfaces correspond to the highest value of the load power factor (0.4), and the upper ones to the lowest (0.1).

The following conclusions are drawn from its form:

1. As $cos\varphi_L$ decreases, the ratio U/U_{OUT} increases;

2. The frequency coefficient ν has a weak effect on the ratio U/U_{OUT}, compared to the influence of the damping coefficient k, as with the increase in ν, the ratio U/U_{OUT} increases.

3. As the damping coefficient k increases, the ratio U/U_{OUT} decreases.

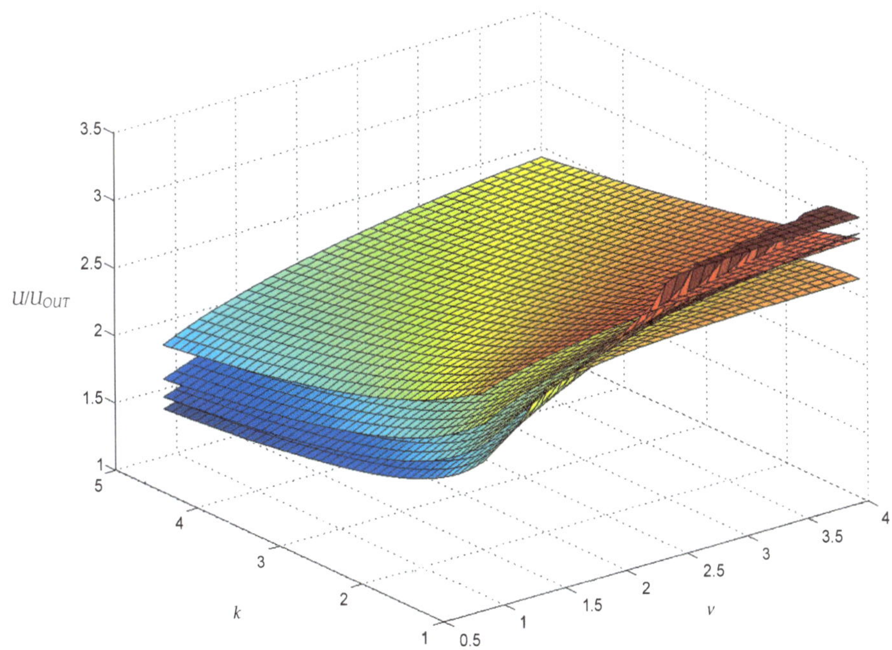

Figure 82. Characteristics of a parallel–series resonant output circuit at a parameter value of s = 0.5. Source: Figure by author.

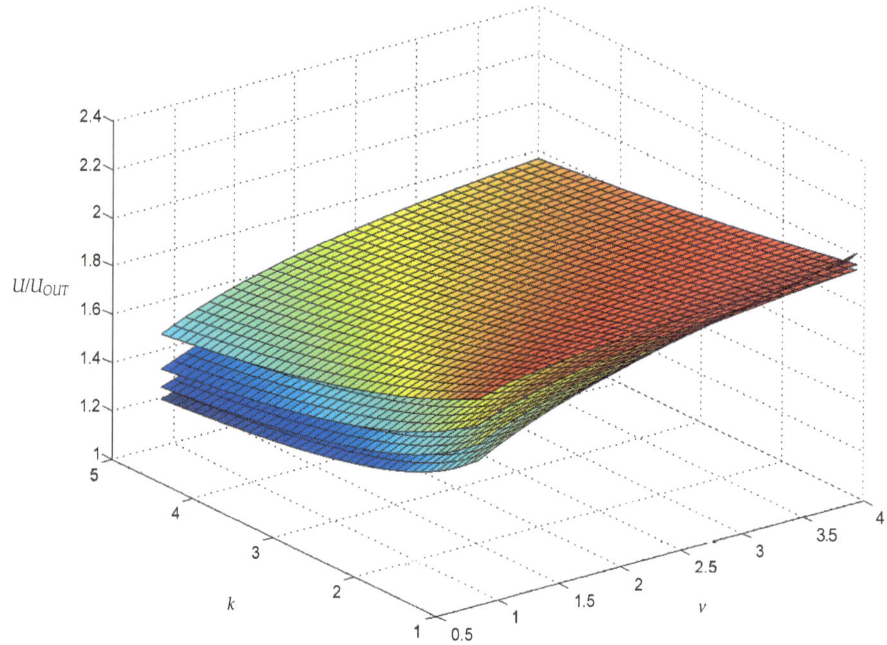

Figure 83. Characteristics of a parallel–series resonant output circuit at a parameter value of $s = 1$. Source: Figure by author.

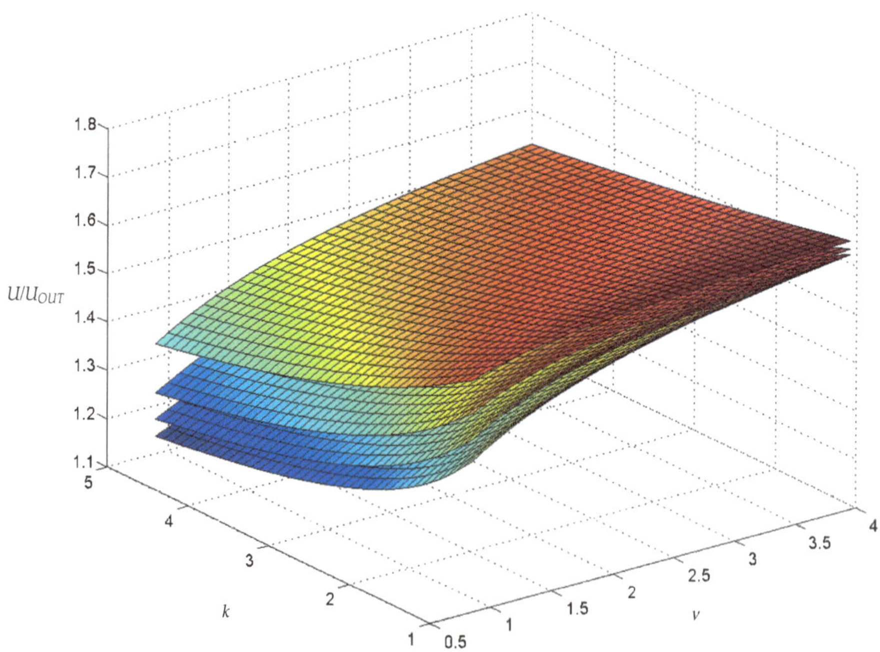

Figure 84. Characteristics of a parallel–series resonant output circuit at a parameter value of $s = 1.5$. Source: Figure by author.

On the other hand, a comparison of the characteristics from the various figures shows that an increase in the coefficient s results in a reduction in the ratio U/U_{OUT}. The plotted characteristics confirm the conclusions drawn in previous studies, namely that the parallel–series resonant output circuit, when operated in the selected mode, is best suited for matching a DC/AC converter to a load requiring a voltage higher than the inverter's output voltage. In other words, this configuration is particularly appropriate for high-power resonant inverters.

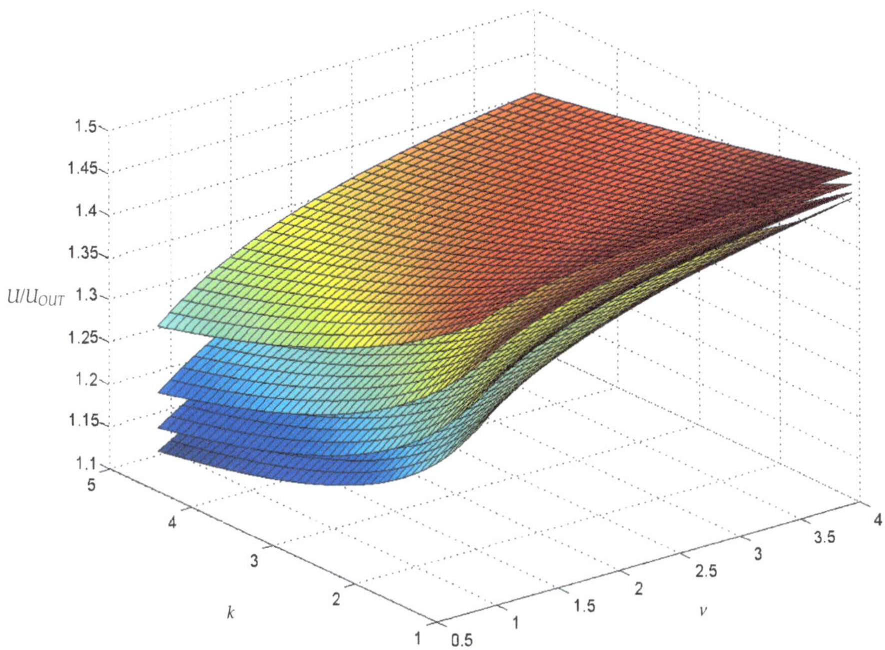

Figure 85. Characteristics of a parallel–series resonant output circuit at a parameter value of $s = 2$. Source: Figure by author.

4.4.3. Series–Parallel–Parallel–Series Resonant Output Circuit

The schematic diagram of a series–parallel–parallel–series resonant output circuit is presented in Figure 86. To configure a series–parallel–parallel–series resonant inverter, this output stage is connected to any of the power circuits shown in Figure 69 at the points marked A and B. The same notation as in the preceding sections is applied. In this configuration, the circuit in Figure 86 can be represented as a combination of a series–parallel output circuit and a parallel–series output circuit. This representation enables the direct application of the results obtained in the previous two sections to determine the relationship $\frac{U}{U_{OUT}} = f(p, s)$.

As with the previously examined output circuits, the relationship between the inverter's output voltage and the load voltage is derived from the balance of active power at the output:

$$\frac{U}{U_{OUT}} = \frac{cos\beta}{cos\gamma}\frac{cos\varphi}{cos\varphi_L}. \tag{153}$$

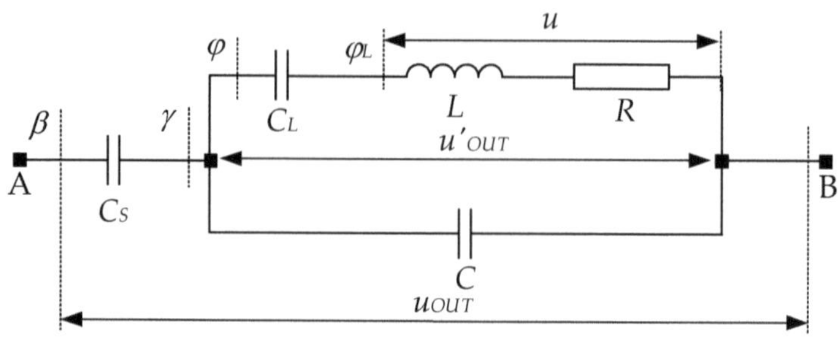

Figure 86. Series–parallel–parallel–series resonant output circuit. Source: Reprinted from [37], used with permission.

To solve Equation (153), it is necessary to determine the unknown angles γ and φ. Since the resulting equations for these variables are of the fourth degree, deriving an analytical dependence becomes complex and labor-intensive. Moreover, the graphical representations of the sought function become multidimensional, making them impractical for rapid engineering design.

For this reason, when designing resonant inverters incorporating a series–parallel–parallel–series resonant output circuit, it is more rational to utilize the results obtained in Sections 4.5.1 and 4.5.2.

4.5. Unified Design Methodologies for Resonant (With and Without Reverse Diodes) and Voltage-Source DC/AC Converters in Different Operating Modes

The analysis of resonant DC/AC converters with and without reverse diodes, as well as voltage-source inverters operating in different modes, combined with the study of the characteristics of the main output circuits used for matching the parameters of the converter and the load, enables the development of comprehensive engineering methodologies for designing an entire class of power electronic devices.

In the design process, the following output specifications are typically provided:

- The output active load power P;
- $cos\varphi_L$ of the load;
- The RMS value of the load voltage U;
- Output frequency $f(\omega)$.

Depending on the type of AC power supply grid and the rectifier topology, the value of the DC supply voltage is known in advance. For the sake of achieving high efficiency and optimal utilization of the semiconductor devices, the damping factor k and the frequency factor ν are selected appropriately. The method of matching between the DC/AC converter and the load is determined according to the output power and load requirements.

As established in Section 4.3, to achieve operation of a resonant inverter (RI) in CCM, close to that of an ideal CSI, the coefficients should be chosen as $k \geq 2.5$ and $\nu \geq 3$. For an RI without reverse diodes operating in soft commutation, a good harmonic composition of the AC circuit current is obtained at $k \geq 1.3$ and $\nu \geq 0.85$. For RIRDs, the quasi-boundary method is effective when $k \geq 1.3$ and $0.85 \leq \nu \leq 1.15$ [33,36,85].

The foundation of the design methodology is the fundamental relationship between the output and input voltages of the RI, which, depending on the operating mode and the type of power circuit, is given by Equations (134) or (135). Since these equations differ only in the numerical coefficients, the generalized form should be applied for design purposes:

$$U_{OUT} = TF\frac{U_d}{cos\beta}, \tag{154}$$

where $TF = \begin{cases} \frac{\pi}{2\sqrt{2}} - \textit{in resonant inverters with hard commutation (CSI)}; \\ \frac{2\sqrt{2}}{\pi} - \textit{in resonant inverters with soft commutation (with and without reverse diodes)}; \end{cases}$

The other key relationship required for design is Equation (133), as it links the coefficients k and ν to the values of the elements in the converter's AC circuit.

According to [17], in a CSI the RMS value of the output voltage—when supplied from a DC source obtained by rectifying a standard three-phase AC grid (without the use of an input grid transformer)—typically lies in the range of 700 to 850 V. In a soft-commutated RI, this value is approximately 25% lower. As previously noted, higher or lower RMS load voltages can be achieved by incorporating an output transformer, by using an inductor with multiple taps, or by employing complex resonant output circuits.

Since the primary objective of this monograph is to present rational design methodologies for DC/AC converters that account for the specific nature of their operation, simplified dependencies will be applied to determine the current and voltage stresses on the semiconductor devices.

From the principles of operation and the analysis of RI, it is known that an estimate of the device loading can be expressed as [5,15,17] and is given by the following:

1. Maximum current through semiconductor devices, I_{max}:

$$I_{max} \approx \begin{cases} I_d - \textit{in resonant inverters with hard commutation (CSI)}; \\ I_d\frac{\pi}{2} - \textit{in resonant inverters with soft commutation (with/without reverse diodes)}; \end{cases}$$

2. Average current through semiconductor devices, I_{av}:

$$I_{av} = \frac{I_d}{2}. \tag{155}$$

3. Maximum voltage on semiconductor devices, U_{max}:

$$U_{max} = \sqrt{2}U_{OUT}. \tag{156}$$

4. Schematic recovery time (for thyristors), t_{qc}:

$$t_{qc} = \frac{\beta}{\omega} = \frac{\beta}{360f}. \tag{157}$$

The design methodology will first be demonstrated for the simplest case—a parallel inverter. Subsequently, other practical configurations for matching the inverter to the load will be examined, including implementations with complex resonant output circuits and series DC/AC converters.

4.5.1. Methodology for Designing Resonant DC/AC Converters with Parallel Load Circuit

A distinctive feature of the parallel DC/AC converter is that the output voltage is identical to the load voltage, and the angles introduced in the analysis satisfy the condition $\beta = \gamma$. This characteristic contributes to the relative simplicity of the circuit and facilitates its design.

1. From Equations (134) or (135) and given the known value of the DC power source voltage U_d together with the RMS value of the load (output) voltage $U = U_{OUT}$, the phase angle β of the AC circuit can be determined:

$$\beta = arccos\left(TF\frac{U_d}{U}\right). \tag{158}$$

2. For the reasons discussed earlier, the value of the detuning coefficient k for the series resonant circuit is selected according to the specific power topology. The magnitude of the detuning coefficient ν is then determined using Formula (133):

$$\nu = \frac{\frac{\pi}{ln\frac{k}{k-1}}\left[1-(1-\rho)\nu^2\right] + \frac{ln\frac{k}{k-1}}{\pi}}{2tg\beta}. \tag{159}$$

In most practical cases involving a parallel DC/AC converter, the coefficient ρ is equal to 1, in which case Equation (159) is modified accordingly:

$$\nu = \frac{\frac{\pi}{ln\frac{k}{k-1}} + \frac{ln\frac{k}{k-1}}{\pi}}{2tg\beta}. \tag{160}$$

3. Determining the natural resonant frequency of the series resonant circuit:

$$\omega_0 = \frac{\omega}{\nu} \tag{161}$$

4. Finding the amount of current consumed by the DC power source: I_d.

The current drawn from the DC power source can be calculated using the results obtained from the analysis of the specific power circuit. However, since the purpose of this work is to propose a convenient and rational engineering methodology for designing resonant inverters, an alternative approach for its determination is suggested here. Under the assumptions made in the analysis—namely, that the inverter operates with unity efficiency—active power is consumed solely by the load. In this case, the equation for the supply current is obtained as follows:

$$I_d = \frac{P}{U_d}. \tag{162}$$

In [35,86], it has been demonstrated that the results obtained using this formula show no significant deviation from those calculated with the exact equation.

5. Determination of the active load resistance of the parallel equivalent load circuit:

$$R_E = \frac{U^2}{P}. \quad (163)$$

6. Finding the components of the series load replacement scheme:

$$R = \frac{R_E}{1 + tg^2\varphi_L}, \; L = \frac{Rtg\varphi_L}{\omega}. \quad (164)$$

7. Determination of the first harmonics of the active component $R_{(1)}$ and the reactive component $X_{(1)}$ of the series equivalent circuit of the inverter's AC circuit (in this case, a parallel load circuit):

$$R_{(1)} = R_E cos^2\beta, \; X_{(1)} = R_E cos\beta sin\beta. \quad (165)$$

8. Calculation of the damping of the series resonant circuit:

$$\delta = \frac{\omega_0}{\pi} ln \frac{k}{k-1}. \quad (166)$$

9. Finding the value of resonant inductance:

$$L_R = \frac{R_{(1)}}{2\delta}. \quad (167)$$

10. Determining the value of the load (parallel) capacity:

$$C = \frac{tg\beta + tg\varphi_L}{\omega R_E}. \quad (168)$$

11. Calculating the maximum voltage of the load capacitor:

$$U_{C_{max}} = U_{max} = \sqrt{2}U. \quad (169)$$

Based on the previously determined current and voltage stresses of the semiconductor devices, and taking into account the operating frequencies, the appropriate power switches are selected. If required, parallel and/or series connection of devices is implemented to meet the design specifications.

4.5.2. Methodology for Designing a Resonant Inverter with a Series–Parallel Output Circuit

The schematic of the series–parallel resonant output circuit is shown in Figure 74. Due to the considerable similarity with the design procedure for the parallel resonant DC/AC converter, only the differences specific to this configuration will be outlined in this section. As in the design of the parallel RI, the value of the output voltage must be specified here, since it is not included in the initial design requirements. Under the operational assumptions made for the series–parallel resonant output circuit, the load voltage is always lower than the output voltage.

1. Determining the detuning coefficient of the parallel load circuit γ:

$$cos\gamma = \frac{U_{OUT}}{U} cos\beta. \tag{170}$$

2. Finding the value of the capacitance of the load (parallel) capacitor:

$$C = \frac{tg\gamma + tg\varphi_L}{\omega R_E}. \tag{171}$$

3. Determining the value of the equivalent capacitor from the series equivalent circuit of the series–parallel resonant circuit C_Σ:

$$C_\Sigma = \frac{1}{L_R(\omega_0^2 + \delta^2)}. \tag{172}$$

4. Calculating the capacitance of a series capacitor C_S:

$$C_S = \frac{C_\Sigma C_{(1)}}{C_{(1)} - C_\Sigma}, \text{ where } C_{(1)} = \frac{1}{\omega X_{(1)}}. \tag{173}$$

5. Determining the maximum voltage of a series capacitor U_{CSmax}:

$$U_{C_{Smax}} = X_{C_{Smax}} I_{(1)max} = \frac{1}{\omega C_S} \frac{U_{max}}{R_{(1)}} cos\gamma \tag{174}$$

To determine the remaining quantities required for the design of the series–parallel resonant inverter, the steps outlined in Section 4.4.2 are followed.

4.5.3. Methodology for Designing a Resonant DC/AC Converter with a Parallel–Series Output Circuit

The schematic of the parallel–series resonant output circuit is presented in Figure 80. In the proposed methodology, only the differences from the design of the basic parallel resonant inverter circuit are emphasized. The selection of the RMS value of the output voltage is governed by the same considerations discussed in the preceding sections.

1. Determining the detuning angle of the series load circuit φ:

$$cos\varphi = \frac{U}{U_{OUT}} cos\varphi_L. \tag{175}$$

As presented in Section 4.4.2, the series resonant circuit comprising R, L and C_L is transformed into an impedance consisting of series-connected active (R_{SER}) and reactive (X_{SER}) components. This impedance is then converted into an equivalent form consisting of parallel-connected active ($1/R_E$) and reactive ($1/R_E$) components.

2. The active load resistance of the parallel equivalent load circuit is determined by:

$$R_E = \frac{U_{OUT}^2}{P}. \tag{176}$$

3. Finding the first harmonics of the active, $R_{(1)}$, and reactive, $X_{(1)}$, components of the series equivalent circuit of the inverter AC circuit:

$$R_{(1)} = R_E cos^2\beta,\ X_{(1)} = R_E\ cos\beta sin\beta. \tag{177}$$

4. Calculating the value of the load (parallel) capacity C:

$$C = \frac{tg\beta + tg\varphi}{\omega R_E}. \tag{178}$$

5. Determining the value of series capacitance C_L:

$$C_L = \frac{1}{\omega R(tg\varphi_L - tg\varphi)}. \tag{179}$$

6. Finding the maximum voltage across a series capacitor U_{CLmax}:

$$U_{C_{Lmax}} = X_{C_{Lmax}} I_{Lmax} = \frac{1}{\omega C_L}\frac{U_{Cmax}}{R} cos\varphi = U_{Cmax}(tg\varphi_L - tg\varphi)cos\varphi. \tag{180}$$

7. Determining the maximum voltage of a parallel capacitor U_{Cmax}:

$$U_{C_{max}} = U_{OUTmax} \tag{181}$$

To calculate the remaining quantities required for the design of the parallel–series resonant DC/AC converter, points from part are used in Section 4.5.1.

4.5.4. Methodology for Designing a Resonant DC/AC Converter with a Series–Parallel–Parallel–Series Circuit

The schematic of the series–parallel–parallel–series resonant output circuit is presented in Figure 86. This configuration is regarded as a combination of a series–parallel and a parallel–series resonant output circuit. Analogous to the design approach applied to the other circuit variants, and taking into account the capability of the series–parallel circuit to reduce, and correspondingly of the parallel–series circuit to increase, the output voltage, the value of U'_{OUT} is selected accordingly. Additionally, the value of the output voltage must be determined. In the proposed methodology, only the differences relative to the design procedure of a parallel resonant DC/AC converter are highlighted.

1. Finding the detuning of the series load circuit φ:

$$cos\varphi = \frac{U}{U'_{OUT}} cos\varphi_L. \tag{182}$$

2. Determining the detuning of the parallel load circuit γ:

$$cos\gamma = \frac{U_{OUT}}{U'_{OUT}} cos\beta. \tag{183}$$

Similarly to the design of the parallel–series resonant inverter, the series load resonant circuit, composed of R, L and C_L, is replaced by an impedance consisting of series-connected active (R_{SER}) and reactive (X_{SER}) components. This impedance is then transformed into an equivalent impedance represented by parallel-connected active ($1/R'_E$) and reactive ($1/X'_E$) components.

3. The active load resistance of the parallel equivalent load circuit R'_E is determined with:

$$R'_E = \frac{(U'_{OUT})^2}{P}. \quad (184)$$

4. Finding the first harmonics of the active, $R_{(1)}$ and reactive, $X_{(1)}$, components of the series equivalent circuit of the parallel–series load circuit:

$$R_{(1)} = R'_E cos^2\gamma,\ X_{(1)} = R'_E cos\gamma sin\gamma. \quad (185)$$

5. Calculation of the value of the load (parallel) capacity C:

$$C = \frac{tg\gamma + tg\varphi}{\omega R'_E}. \quad (186)$$

6. Determining the value of the series capacitance associated with the load C_L:

$$C_L = \frac{1}{\omega R(tg\varphi_L - tg\varphi)}. \quad (187)$$

7. Finding the value of the equivalent capacitance from the series equivalent circuit of the output circuit C_Σ:

$$C_\Sigma = \frac{1}{L_R(\omega_0^2 + \delta^2)}. \quad (188)$$

8. Finding the capacitance of a series capacitor C_S:

$$C_S = \frac{C_\Sigma C_{(1)}}{C_{(1)} - C_\Sigma},\ \text{where } C_{(1)} = \frac{1}{\omega X_{(1)}}. \quad (189)$$

9. Determining the maximum voltage of the series capacitor connected to the load U_{CLmax}:

$$U_{C_{Lmax}} = X_{C_L} I_{Lmax} = \frac{1}{\omega C_L} \frac{U'_{OUTmax}}{R} cos\varphi = U'_{OUTmax}(tg\varphi_L - tg\varphi)cos\varphi. \quad (190)$$

10. Calculating the maximum voltage of a parallel capacitor U_{Cmax}:

$$U_{C_{max}} = U'_{OUTmax}. \quad (191)$$

11. Determining the maximum voltage of a series capacitor U_{CSmax}:

$$U_{C_{Smax}} = X_{C_S} I_{(1)max} = \frac{1}{\omega C_S} \frac{U'_{OUTmax}}{R_{(1)}} cos\gamma. \tag{192}$$

To calculate the remaining quantities required for the design of a series–parallel–parallel–series resonant inverter, the steps outlined in Section 4.5.1 are followed.

4.5.5. Methodology for Designing a DC/AC Converter with a Series Resonant Circuit

The series resonant DC/AC converter can be implemented either with or without reverse diodes. The design procedure is performed in the following sequence. The coefficients k and ν are selected based on the considerations discussed in the section on the unified approach to the analysis of resonant DC/AC converters. Since the voltage across the series resonant circuit is generated by the commutation of each pair of semiconductor devices, the relationship between the RMS value of the first harmonic of the output voltage and the voltage of the input DC power source is given by [61]:

$$U_{OUT} = \frac{2\sqrt{2}}{\pi} U_d \tag{193}$$

If the resulting value of the device's output voltage does not match the requirements of the design specification, an output transformer must be used. Here is a brief summary of the design steps:

1. The magnitude of the load resistance R is determined by:

$$R = \frac{U_{OUT}^2}{P} \tag{194}$$

2. Finding the natural frequency of the series resonant circuit ω_0:

$$\omega_0 = \frac{\omega}{\nu} \tag{195}$$

3. Determining the damping of the series resonant circuit δ:

$$\delta = \frac{\omega_0}{\pi} ln \frac{k}{k-1} \tag{196}$$

4. Calculating the value of resonant inductance L_R:

$$L_R = \frac{R}{2\delta} \tag{197}$$

5. Determining the value of resonant capacitance C_R:

$$C_R = \frac{1}{L_R(\omega_0^2 + \delta^2)} \tag{198}$$

6. The calculation of the magnitude of the average current through the devices, I_{av}, the current consumed by the DC power source, I_d, and the maximum current in the AC circuit: I_{max}, is performed as in the case of a parallel resonant DC/AC converter with soft commutation.

7. Determining the maximum voltage value on the resonant capacitor, $U_{C_{Rmax}}$:

$$U_{C_{Rmax}} = X_{CR} I_{max} = \frac{1}{\omega C_R} I_{max} \tag{199}$$

Based on the previously determined current and voltage stresses of the semiconductor devices, and taking into account the operating frequencies, the appropriate power switches are selected. If required, parallel and/or series connection of devices is employed to meet the design specifications.

4.6. Voltage-Source Inverter Design Methodology

Typically, when designing a VSI, the following parameters are set: output active power P/output apparent power S, the RMS value of the voltage on the load U, the operating frequency f and the power factor of the load $cos\varphi_L$. Below is an example sequence for determining the most important parameters necessary for the design of the devices under consideration.

1. Based on the relationship between the power factor of the load and its parameters, the value of the quasi-frequency coefficient is determined:

$$\nu_A = 2\sqrt{\frac{1 - cos^2\varphi_L}{cos^2\varphi_L}} \tag{200}$$

2. At a selected control frequency, the damping of the series circuit is found:

$$\delta = \Omega_0 = \frac{\omega}{\nu_A} = \frac{2\pi f}{\nu_A} \tag{201}$$

3. From Equation (80), with value of the quasi-frequency coefficient already found, the required value of input voltage is determined.

$$U_d = \frac{U}{D_N G_N}. \tag{202}$$

If the obtained value is not compatible with the DC power source, a DC–DC converter should be used at the converter input, or an inverter transformer should be implemented at the output.

4. The active component of the load resistance is determined based on the active output power:

$$R = \frac{U^2}{P}. \tag{203}$$

5. The reactive component of the load resistance is respectively:

$$L = \frac{Rtg\varphi_L}{2\pi f} \tag{204}$$

Equations (72), (74), (76) and (77) define, respectively, the maximum instantaneous current through the transistors, the average current through the transistors, the average input current drawn from the DC power source, and the average current flowing through the reverse diodes.

4.7. Examples of Application of Design Methodologies

To demonstrate the suitability of the presented methodologies for designing various types of DC/AC converters with and without reverse diodes, computer simulations of transient and steady states have been performed using the LTspice simulator, as well as models developed by the author in MATLAB/Simulink. The presented methodologies have been used in the educational process conducted by the Department of Power Electronics at the Technical University of Sofia.

4.7.1. Example of Designing a Parallel Resonant DC/AC Converter, Operating in Soft Commutation Mode (RI)

Initial data for the design of a parallel DC/AC converter without reverse diodes, operating in soft commutation mode are as follows (scheme in Figure 69a):

- Output active power of the load $P = 100$ kW;
- Load power factor $cos\varphi_L = 0.15$;
- RMS value of the load voltage $U = 850$ V;
- Output frequency $f = 4000$ Hz.

In accordance with the design guidelines, the values of the coefficients $k = 1.5$ and $\nu = 1$ were selected. As a result of the design, the following values of the circuit elements of the inverter were determined: $R = 0.163\ \Omega$, $L = 42.72\ \mu$H, $C = 45.02\ \mu$F and $L_R = 115.3\ \mu$H. For the design purposes, the DC power supply voltage value $U_d = 500$ V was chosen.

Figure 87 presents the results of numerical experiments performed with the model of a parallel thyristor resonant inverter, implemented in MATLAB/Simulink (shown in Figure 52). From top to bottom, the diagrams display: the output voltage (i.e., the voltage across the parallel capacitor C) (u_{OUT}), the input current (i_d) flowing through the resonant inductor, and the current through the load circuit (i).

Figure 88 presents simulation results from the study of the designed converter—a parallel thyristor resonant DC/AC converter. From top to bottom, the waveforms illustrate: the input current (i_L) flowing through the inductance, the output voltage (u_{OUT}) across the parallel capacitor C, and the voltage across thyristor VS_1 (U_{VS1}). The simulation model, by which these results were obtained, is described in [36] and was implemented in LTspice. Since the simulator does not include a native thyristor model, the thyristor was emulated by a series connection of a bipolar NPN transistor and a semiconductor diode, forming a group that accurately reproduces the actual switching behavior of a thyristor.

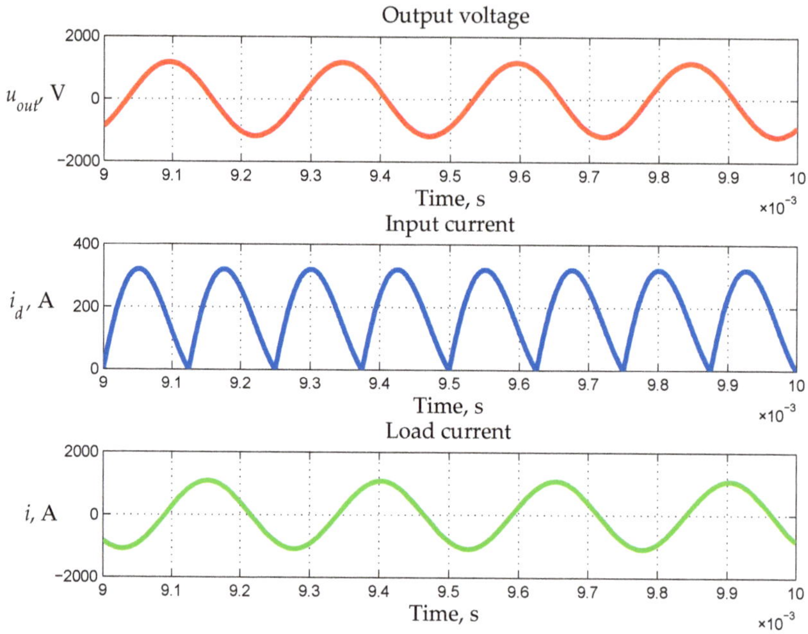

Figure 87. Timing diagrams from numerical experiments with a model of the parallel resonant DC/AC converter in steady state, as follows: output voltage—u_{OUT}; input current—i_d; current through the load circuit—i. Source: Figure by author.

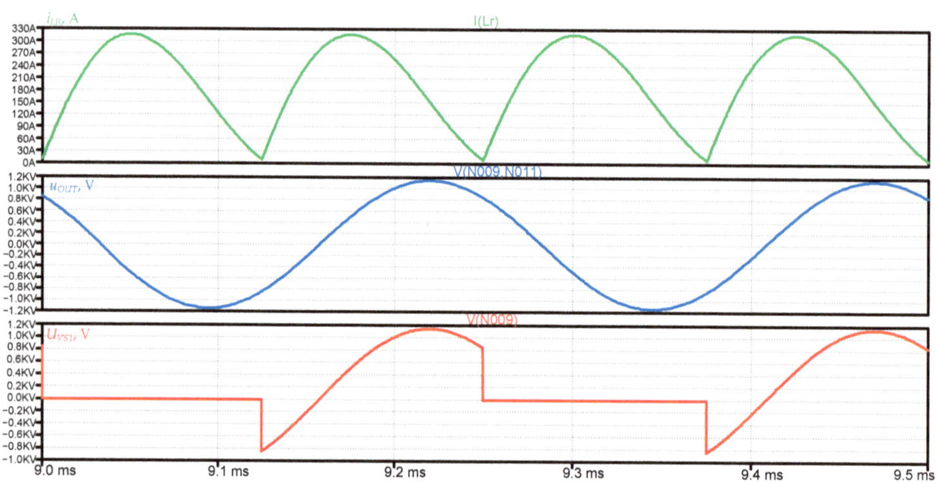

Figure 88. Timing diagrams from simulations of the parallel resonant DC/AC coverter in steady state: input current—i_{LR}, output voltage—u_{OUT} and voltage on thyristor VS_1—U_{VS1}. Source: Reprinted from [37], used with permission.

Table 2 compares the results obtained through the unified analysis [33,36] and the design methodology based on it for a parallel resonant DC/AC converter, with those derived from simulation and numerical experiments using the developed model. The simulation results were obtained through LTspice and verified using numerical calculations in mathematical software, applying the basic computational functions to extract the relevant electrical quantities. The comparison shows that the values from both approaches match within the acceptable range of engineering tolerance, which confirms the accuracy and practical applicability of the proposed design methodology.

Table 2. Initial data and obtained simulation and modeling results for a full-bridge thyristor parallel resonant DC/AC converter.

Initial Data	LTspice	Matlab/Simulink
U_{OUT}, V = 850	814.47	832.39
I_d, A = 200	195.04	203.56
I_{max}, A = 314.16	316.85	319.6
t_q, µs = 40.33	41	41.5

Source: Reprinted from [37], used with permission.

The data presented in Table 2 show that for all values determined during design, the error is less than 5%.

4.7.2. Example of Designing a Parallel DC/AC Converter, Operating in Hard Commutation Mode (CSI)

Initial data for the design of a parallel resonant DC/AC converter without reverse diodes, operating in hard commutation mode are as follows (scheme in Figure 69a):

- Output active power of the load P = 100 kW;
- Load power factor $cos\varphi_L$ = 0.15;
- RMS value of the load voltage U = 750 V;
- Output frequency f = 2400 Hz.

According to the design guidelines, the values of the coefficients k = 2.5 and ν = 3 were selected. As a result of the design, the following values of the circuit elements of the inverter were determined: R = 0.12656 Ω, L = 55.319 µH, C = 88.406 µF and L_d = L_R = 2.187 mH. For the design purposes, the DC power supply voltage value U_d = 500 V was chosen.

Figure 89 presents the results of numerical experiments performed with the model of a parallel thyristor CSI, implemented in MATLAB/Simulink and shown in Figure 52. From top to bottom, the waveforms represent the output voltage u_{OUT} (voltage across the parallel capacitor C), the input current i_d (current through the inductor), and the current through the load circuit i. These results illustrate the dynamic behavior of the converter in steady-state operation, confirming the expected phase relationships and amplitude levels derived from the analytical design methodology.

Table 3 presents a comparison between the results calculated using the unified analysis methodology [33,36] and the corresponding CSI design procedure based on it, and those obtained from simulation and numerical experiments. The simulation and experimental values were determined through the basic computational functions of the computer simulator and supporting mathematical software. This comparison validates the accuracy and applicability of the proposed design approach by demonstrating a close match between analytical predictions and simulated behavior.

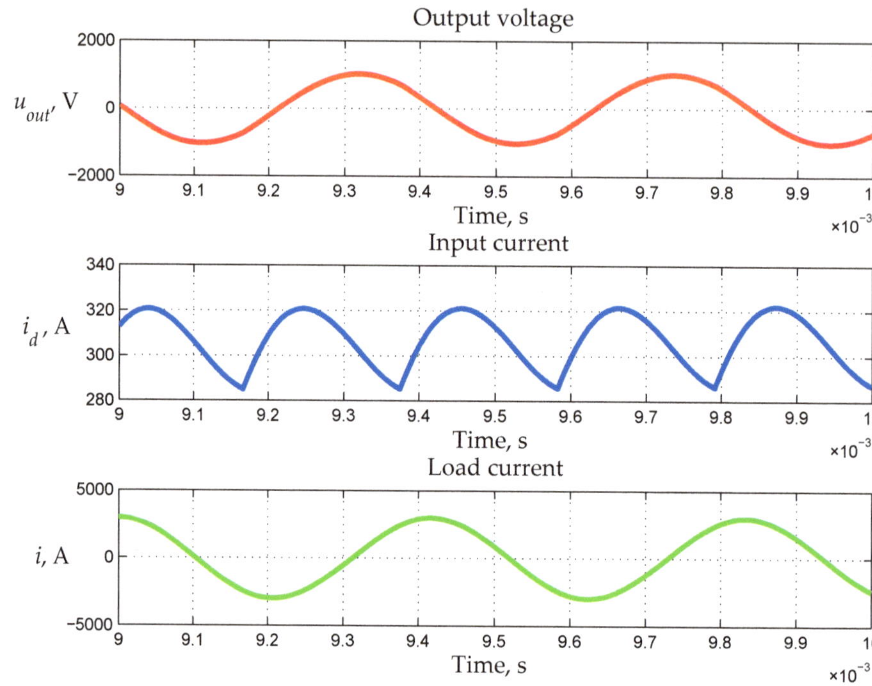

Figure 89. Timing diagrams from numerical experiments with a model of the parallel CSI in steady state, as follows: output voltage—u; input current—i_d; current through the load circuit—i. Source: Figure by author.

Table 3. Initial data and obtained simulation and modeling results for a full-bridge thyristor parallel current-source inverter.

Initial Data	LTspice	Matlab/Simulink
U_{OUT}, V = 750	724.5	725.25
I_d, A = 200	195	195
t_q, μs = 48.87	47	47.5

Source: Table by author.

The data presented in Table 3 show that for all values determined during the design, the error is less than 5%, which validates the methodology used for the design of parallel current-source inverters.

4.7.3. Example of Designing a Series–Parallel DC/AC Converter, Operating in Hard Commutation Mode (CSI)

The following calculation example is based on the design of a thyristor resonant series–parallel DC/AC converter, operating in hard commutation mode (CSI).

The initial data for the design of a full-bridge thyristor series–parallel current-source inverter are as follows:

- Active output power $P = 100$ kW;
- Load power factor $cos\varphi_L = 0.15$;

- RMS value of the load voltage $U = 750$ V;
- RMS value of the output (inverter) voltage $U_{OUT} = 800$ V;
- Output frequency $f = 2400$ Hz.

According to the design guidelines, the values of the coefficients $k = 2.5$ and $\nu = 3$ were selected. As a result of the design, the following values of the circuit elements were determined: $R = 0.127\ \Omega$, $L = 55.32\ \mu$H, $C = 88.42\ \mu$F, $C_S = 166.6\ \mu$F and $L_R = 1.909$ mH. For the design purposes, the voltage value of the DC power source $U_d = 500$ V was selected.

Figure 90 shows the simulation results of the study of the designed DC/AC converter: series–parallel CSI. The figure presents sequentially from top to bottom the input current, i_{LR}; the output voltage, U_{OUT}; the voltage on the load, U; and the voltage on the thyristor VS_1, U_{VS1}. The simulation model of the CSI, through which the presented results were obtained, is presented in the corresponding section of this book.

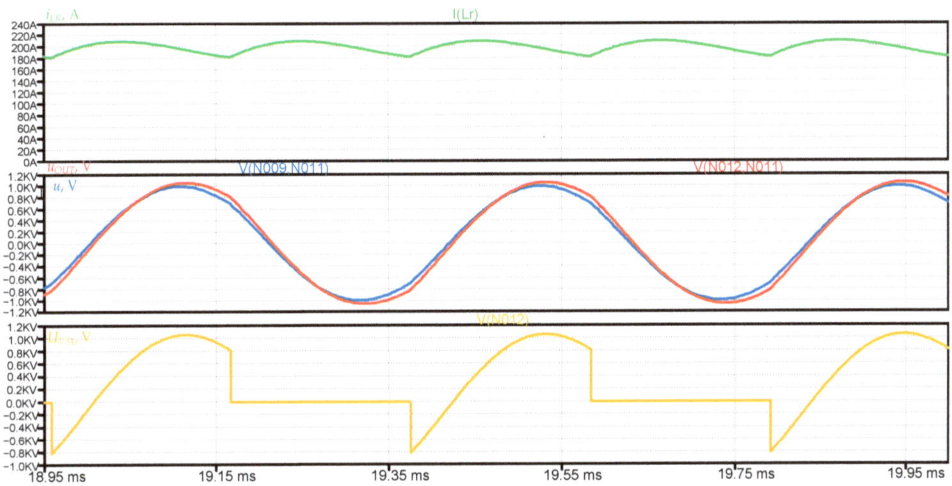

Figure 90. Timing diagrams describing the operation of the series–parallel CSI: input current—i_{LR}; output voltage—u_{OUT}; load voltage u; voltage on the thyristor VS_1, U_{VS1}. Source: Figure by author.

Table 4 compares the results obtained using the unified analysis and the corresponding design methodology for a full-bridge series–parallel current-source inverter with those derived from simulations and numerical experiments with the model.

Table 4. Initial data and obtained results from simulation and numerical experiment for a thyristor full-bridge series–parallel current-source inverter.

Initial Data	LTspice	Matlab/Simulink
U_{OUT}, V = 800	776.8	775.25
U, V = 750	728	725.6
I_d, A = 200	198.39	196
t_q, μs = 53.33	51	50.6

Source: Table by author.

The data presented in Table 4 demonstrate that, for all design-determined values, the deviation between calculated and simulated results does not exceed 5%.

The results confirm the accuracy of the unified analysis and the derived design methodology for a series–parallel CSI. The correspondence between the calculated values and the simulated waveforms demonstrates that the proposed approach ensures correct prediction of the operating parameters. This validates the applicability of the methodology for practical engineering design and for selecting appropriate component values in real-world implementations.

4.7.4. Example of Designing a Thyristor Parallel–Series DC/AC Converter, Operating in Hard Commutation Mode (CSI)

The following calculation example is based on the design of a thyristor parallel–series CSI. The initial data for the design of a full-bridge thyristor parallel–series CSI are:

- Active output power $P = 250$ kW;
- Load power factor $cos\varphi_L = 0.09$;
- RMS value of the load voltage $U = 1550$ V;
- RMS value of the output (inverter) voltage $U_{OUT} = 750$ V;
- Output frequency $f = 2400$ Hz.

When designing in accordance with the given task, an operating mode was selected in which the control frequency exceeds the resonant frequency of the equivalent series resonant circuit. In this case, the detuning coefficient has a value of $\nu = 3.5$, while the coefficient k is set to $k = 2.5$.

As a result of the design, the following values of the circuit elements of the device were determined: $R = 0.073\ \Omega$, $L = 53.5\ \mu$H, $C = 187.6\ \mu$F, $C_L = 162.4\ \mu$F and input/resonant inductance $L_R = 885.6\ \mu$H. For the design purposes, a value of the input DC power supply of 500 V was selected.

Figure 91 presents the simulation results from the study of the designed DC/AC converter: parallel–series CSI. From top to bottom, the plots show the input current (i_{LR}), the output voltage (U_{OUT}), the load voltage (U), and the voltage across thyristor VS_1 (U_{VS1}). The simulation model used to obtain these results is described in [37].

Table 5 compares the results calculated using the unified analysis and its associated design methodology for a full-bridge parallel–series CSI with those obtained from the simulator and numerical experiments on the model.

Table 5. Initial data and obtained results from simulation and numerical experiment for a full-bridge thyristor parallel–series current-source inverter.

Initial Data	LTspice	Matlab/Simulink
U_{OUT}, V = 750	723.25	722.77
U, V = 1500	1440.7	1439.8
I_d, A = 500	482	481
t_q, μs = 48.66	48	48

Source: Reprinted from [37], used with permission.

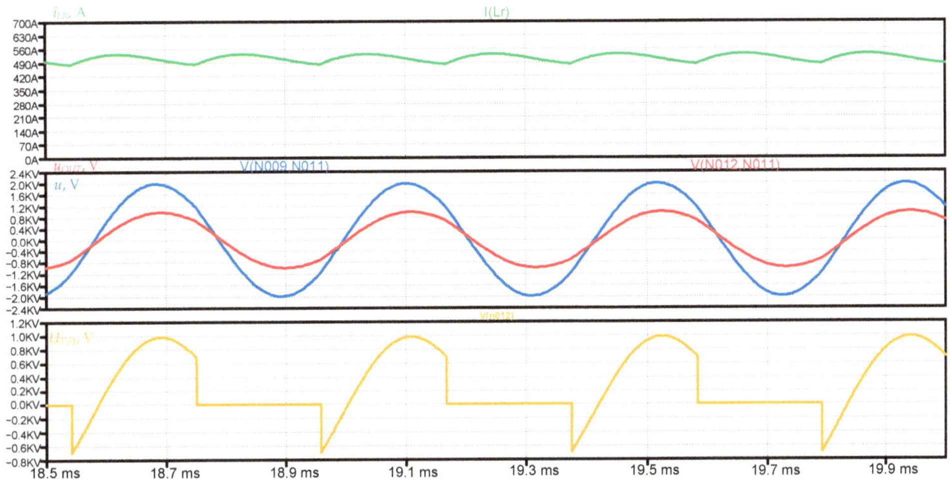

Figure 91. Timing diagrams describing the operation of a parallel–series current-source inverter: input current—i_{LR}; output voltage—u_{OUT}; load voltage, u; voltage on the thyristor VS1, U_{VS1}. Source: Reprinted from [37], used with permission.

The data presented in Table 5 indicate that, for all parameters determined during the design process, the deviation between the calculated values (based on the unified analysis and the proposed methodology) and those obtained from simulation and numerical experiments is less than 4%. This low level of discrepancy demonstrates the high accuracy and reliability of the adopted design approach, confirming its suitability for practical engineering applications of full-bridge parallel–series CSI converters.

4.7.5. Example of Designing a Series Transistor Resonant Inverter with Reverse Diodes

The following computational example is based on the design of a transistor resonant inverter with reverse diodes (scheme in Figure 69d). This example is examined in detail in the context of the unified approach for the analysis of DC/AC converters presented in [33,36].

The initial parameters for the design of the full-bridge transistor resonant inverter with reverse diodes are as follows:

- Active power in the load P = 10 kW;
- Load power factor $cos\varphi_L$ = 1;
- RMS value of the load voltage U_{OUT} = 270 V;
- Output frequency f = 50,000 Hz.

An operating mode is selected in which the control frequency is set higher than the resonant frequency of the series resonant circuit, ensuring stable operation and improved commutation conditions. For this mode, the detuning coefficient is chosen as ν = 1.1, which allows for maintaining efficient energy transfer while avoiding excessive reactive power circulation and minimizing switching losses in the semiconductor devices.

As a result of the design, the following values of the DC/AC converter elements were determined: $R = 7.29\ \Omega$, $L = 36.496\ \mu H$ and $C = 299.32$ nF. For the purposes of the design, a value of the DC power source of 300 V was selected.

Figure 92 presents the results from a computer simulation of the designed DC/AC converter: a series resonant inverter with reverse diodes. From top to bottom, the waveforms illustrate the current drawn from the DC power source (i_d), formed by the sequential conduction of transistors and reverse diodes, and the output voltage (u_{OUT}).

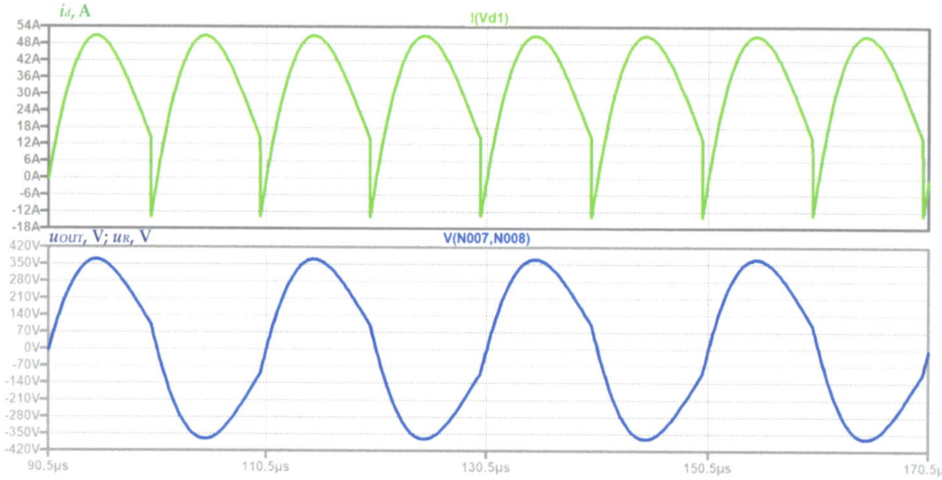

Figure 92. Timing diagrams describing the operation of an RIRD: input current—i_d; output voltage—u_{OUT}. Source: Reprinted from [37], used with permission.

A distinctive feature of this power topology is that, during the conduction intervals of the reverse diodes, part of the stored energy in the AC circuit is returned to the DC source. This phenomenon manifests as negative segments in the waveform of the input current, corresponding to the intervals of energy feedback from the AC circuit to the DC supply. The simulation model used to obtain these results is described in detail in [37].

Table 6 presents a comparison between the results obtained through the unified analytical approach for a full-bridge resonant inverter with reverse diodes and those derived from computer simulation and numerical experiments with the corresponding model. The analysis of the data demonstrates a high degree of agreement between the calculated and simulated values, with deviations remaining within acceptable engineering tolerances. This confirms the validity and practical applicability of the unified analytical method for accurate prediction of inverter performance parameters.

Table 6. Initial data and obtained results from simulation and numerical experiment for a transistor resonant inverter with reverse diodes.

Initial Data	LTspice	Matlab/Simulink
U_{OUT}, V = 270	268.73	267.12
I_d, A = 33.333	33.097	32.45

Source: Reprinted from [37], used with permission.

The data given in Table 6 show that for the values of the quantities determined during the design, the error is less than 5%.

4.7.6. Example of Designing a Parallel Resonant Inverter with Reverse Diodes

Initial data for the design of a parallel transistor resonant DC/AC converter with reverse diodes, operating in a mode with a control frequency above the resonant one are (scheme shown in Figure 62):

- Output active power of the load $P = 10$ kW;
- Load power factor $cos\varphi_L = 0.2$;
- RMS value of the load voltage $U = 700$ V;
- Output frequency $f = 10{,}000$ Hz.

In accordance with the design guidelines, the values of the coefficients $k = 1.35$ and $\nu = 1.15$ were selected. As a result of the design, the following values of the circuit elements of the inverter were determined: $R = 0.98\ \Omega$, $L = 71.46\ \mu$H, $C = 3.961\ \mu$F and $L_R = 214.1\ \mu$H. For the design purposes, the voltage value of the DC power source $U_d = 500$ V was chosen.

Figure 93 presents the simulation results obtained from the LTspice analysis of the designed converter: a parallel transistor RIRD. Sequentially from top to bottom, the waveforms illustrate the current through the resonant inductance (i_{LR}), the output voltage across the parallel capacitor (u_{OUT}), and the voltage across transistor VS_1 (U_{VS1}). The visualized results provide clear insight into the dynamic behavior of the inverter under the specified operating conditions, allowing for verification of the theoretical predictions and evaluation of voltage and current stresses on the key semiconductor components.

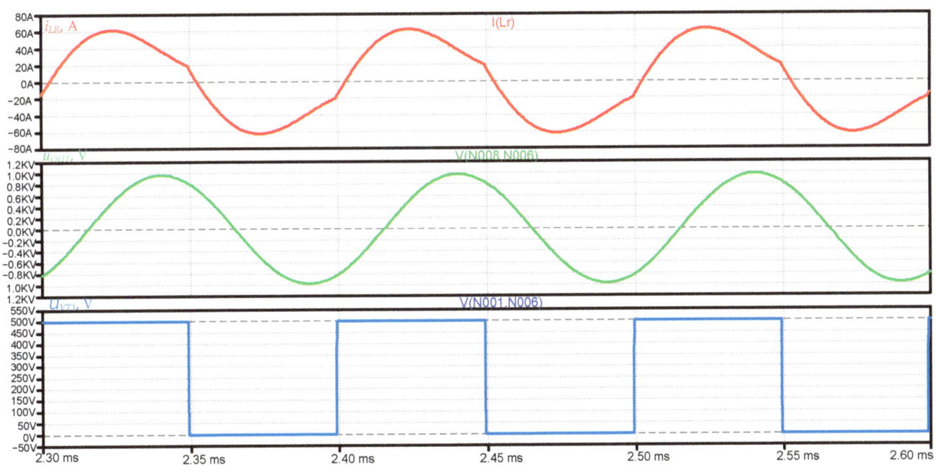

Figure 93. Timing diagrams from simulations of the full-bridge parallel resonant inverter with reverse diodes in steady state: current through the resonant inductance—i_{LR}; output voltage—u_{OUT}; voltage on the transistor VT_1, U_{VT1}. Source: Figure by author.

Table 7 presents a comparison between the results calculated using the unified analysis and the corresponding methodology for designing the RIRD, and those obtained from both the simulator and numerical experiments with the model. The simulation-based values were derived through the standard computational functions available in the computer simulator, while the numerical experiment results were processed using dedicated mathematical software. This comparison enables a direct evaluation of the methodology's accuracy and reliability in predicting the operating parameters of the RIRD.

Table 7. Initial data and obtained results from simulation and modeling for a full-bridge transistor parallel resonant inverter with reverse diodes.

Initial Data	LTspice	Matlab/Simulink
U_{OUT}, V = 700	689.53	690.21
I_d, A = 40	39.42	39.61
I_{max}, A = 62.8	61.9	62

Source: Table by author.

The data presented in Table 7 show that, for all values determined during the design, the deviation does not exceed 3%, confirming the high accuracy and applicability of the proposed methodology.

4.7.7. Example of Designing a Series–Parallel Transistor Resonant DC/AC Converter with Reverse Diodes

The following calculation example is based on the design of a transistor resonant DC/AC converter with reverse diodes (scheme in Figure 69d with a series–parallel output circuit).

The initial data for the design of a full-bridge transistor resonant inverter with reverse diodes are as follows:

- Active power in the load $P = 50$ kW;
- Load power factor $cos\varphi_L = 0.15$;
- RMS value of the output (inverter) voltage $U_{OUT} = 800$ V;
- RMS value of the load voltage $U = 750$ V;
- Output frequency $f = 10{,}000$ Hz.

An operating mode with a control frequency greater than the resonant frequency of the series resonant circuit is selected—the value of the detuning factor $\nu = 1.15$—and the detuning factor is determined using the transfer characteristic of the soft-switched RI (135) and is $k = 1.55$. As a result of the design, the following values of the circuit elements were determined: $R = 0.25312\ \Omega$, $L = 26.55\ \mu$H, $C = 11.218\ \mu$F, $C_S = 28.888\ \mu$F and $L_R = 113.11\ \mu$H. For the design purposes, the voltage value of the DC power source $U_d = 500$ V was selected.

Figure 94 presents the simulation results from the study of the designed DC/AC converter: a series–parallel RIRD. Displayed sequentially from top to bottom are: the current through the resonant inductance (i_{LR}), the output voltage (u_{OUT}), the voltage across the load (u), and the voltage across the transistor VS_1 (U_{VS1}).

The simulation model of the RIRD, by means of which these results were obtained, is provided in the corresponding section of this book. The presented waveforms clearly illustrate the dynamic behavior of the converter under the selected operating mode, validating both the design methodology and the analytical predictions.

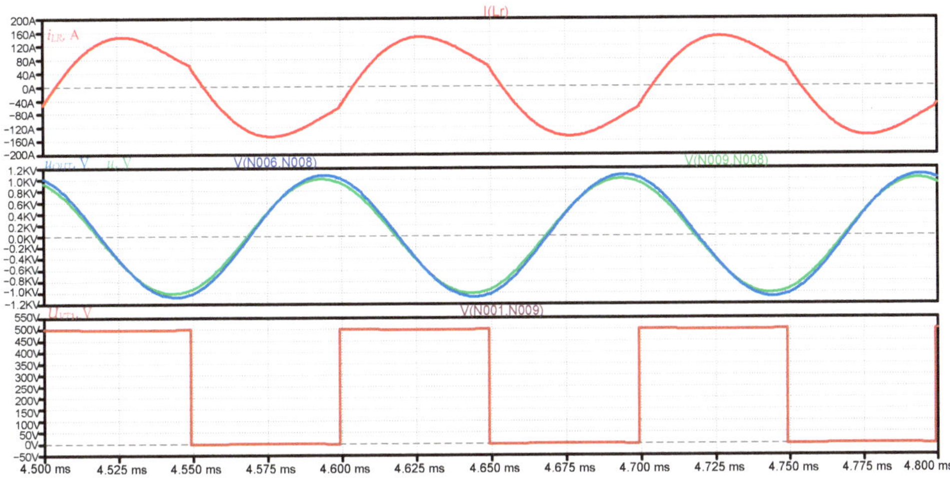

Figure 94. Timing diagrams from simulations of a series–parallel RIRD in steady-state mode: current through the resonant inductance—i_{LR}; output voltage—u_{OUT}; voltage across the load, u; voltage across the transistor, VT_1, U_{VT1}. Source: Figure by author.

The data presented in Table 8 compare the results obtained using the unified analysis of a full-bridge RIRD to those derived from simulator studies and numerical experiments with the corresponding model. The simulation-based values were determined using the built-in computational functions of the circuit simulator and dedicated mathematical software. The comparison demonstrates a high level of agreement between the analytical and simulated results, thereby validating both the accuracy of the unified analysis and the reliability of the applied design methodology.

Table 8. Initial data and obtained results from simulation and numerical experiment for a series–parallel transistor resonant inverter with reverse diodes.

Initial Data	LTspice	Matlab/Simulink
U_{OUT}, V = 800	760.25	768. 5
U, V = 750	716.28	718.2
I_d, A = 100	94.057	95.95

Source: Table by author.

The data given in Table 8 show that for the values of the quantities determined during the design, the error is less than 6%.

4.7.8. Example of Designing a Parallel–Series Resonant DC/AC Converter with Reverse Diodes

The presented computational example is based on the design of a parallel–series RIRD. The input data for the design of a bridge parallel–series inverter with reverse diodes are as follows:

- Active output power $P = 50$ kW;
- Load power factor $cos\varphi_L = 0.2$;
- RMS value of the load voltage $U = 1500$ V;

- RMS value of the output (inverter) voltage U_{OUT} = 760 V;
- Output frequency f = 5000 Hz.

When designing in accordance with the task, an operating mode with a control frequency lower than the resonant frequency of the equivalent series resonant circuit was selected—the value of the detuning coefficient $\nu = 0.9$, and the coefficient k has a value of $k = 1.25$. Thus, the circuit can be implemented with both transistors and thyristors.

As a result of the design, the following values of the circuit elements of the device were determined: $R = 1.8\ \Omega$, $L = 280.7\ \mu$H, $C = 10.19\ \mu$F, $C_L = 6.877\ \mu$F and resonant inductance $L_R = 112.4\ \mu$H. For the design purposes, a value of the input DC power supply of 500 V was selected.

Figure 95 presents the simulation results from the study of the designed DC/AC converter: a parallel–series RIRD. From top to bottom, the plots display: the current through the resonant inductance (i_{Lr}), the output voltage (u_{out}), the voltage across the load (u), and the voltage on the transistor VS_1 (U_{VS1}). The simulation model used to obtain these results is described in the corresponding section of this book.

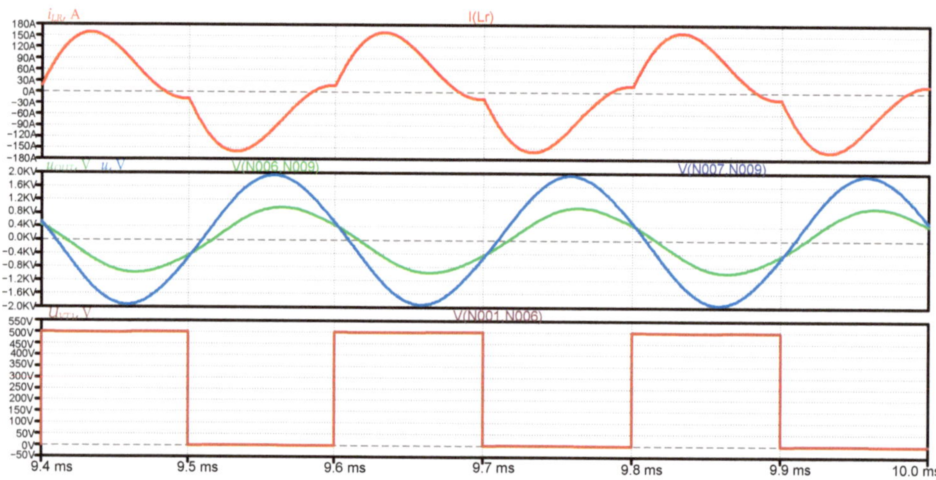

Figure 95. Timing diagrams describing the operation of a parallel–series RIOD: current through the resonant inductance—i_{LR}; output voltage—u_{OUT}; load voltage, u; voltage on the transistor VS_1, u_{VS1}. Source: Figure by author.

Table 9 compares the results calculated using the unified analysis and the associated design methodology for a full-bridge parallel–series RIRD with those obtained from simulator studies and numerical experiments with the model. The simulation-based values were derived using the core computational functions of the circuit simulator and mathematical software. The comparison confirms the close agreement between the analytical and simulated outcomes, thus validating the robustness and applicability of the proposed methodology for practical engineering design.

The data presented in Table 9 indicate that, for all design-determined values, the deviation between the analytical results and those obtained from simulation and numerical experiments does not exceed 5%. This consistently low margin of error confirms the high accuracy and reliability of the proposed unified analysis and its corresponding design methodology when applied to full-bridge parallel–series RIRD configurations.

Table 9. Initial data and obtained results from simulation and numerical experiment for a full-bridge parallel–series RIRD.

Initial Data	LTspice	Matlab/Simulink
U_{OUT}, V = 760	738.47	742.57
U, V = 1500	1461	1485.2
I_d, A = 100	101.91	101.9

Source: Table by author.

4.7.9. Example of Designing a Voltage Source Inverter

The validity of the proposed analysis and the design methodology for the VSI, developed on its basis, was verified through the design of a computational example and the creation of a corresponding model of a full-bridge inverter (Figure 17). The design process was performed using the following input parameters:

- Operating frequency 500 Hz;
- Load power factor $cos\varphi_L = 0.303$;
- RMS value of the first harmonic of the output voltage $U = 100$ V;
- Total output power $S = 1000$ VA.

As a result of the design, the following values of the circuit elements were obtained: load resistance $R = 3.033\ \Omega$ and load inductance $L = 3.033$ mH. From Equation (202) the required value of the input voltage was determined, resulting in a DC power source voltage of $U_d = 111$ V.

Table 10 compares the results obtained from the design of the selected circuit and those obtained from modeling (MATLAB/Simulink) and computer simulations (LTspice). The following notations are used: I_{max}—maximum value of the current through the transistors; I_d—average value of the current consumed by the DC power source; I_{av}—average value of the current through the transistors; I_{dav}—average value of the current through the diodes; I—RMS value of the current through the load; U—RMS value of the output voltage.

Table 10. Values of quantities obtained from the design of a voltage source inverter and the results of modeling and computer simulations.

Initial Data	LTspice	Matlab/Simulink
I_{max}, A = 16.897	16.902	16.91
I_d, A = 2.769	2.771	2.775
I_{av}, A = 2.888	2.89	2.9
I_{dav}, A = 1.504	1.505	1.504
I, A = 10.064	10.064	10.065
U, V = 100	99.99	100

Source: Table by author.

From the data presented in Table 10, it is evident that the proposed design methodology achieves very high accuracy, as the maximum deviation between the calculated values and those obtained from simulations and modeling is less than 0.5%. These discrepancies are primarily attributed to rounding during the calculation procedures and differences in the verification models. Considering that, in the design of power electronic devices, component parameters typically have tolerances of around or above 20%, the presented methodology delivers excellent precision and reliability.

4.8. Conclusions and Key Findings

The conducted design studies, supported by computer simulations and numerical experiments, show that the deviations between calculated and simulated results do not exceed 6%. This accuracy is satisfactory, considering that in practice capacitors are manufactured only in discrete standard values, which makes it impossible to exactly match the theoretical parameters determined by the design methodology. Additionally, passive components in power electronic devices—particularly inductances and capacitances—exhibit substantial manufacturing tolerances, typically 15–30%, and their actual parameters can vary significantly depending on the operating conditions.

In this context, the primary objective of the work has been to develop a simplified, systematic, and rational methodology for designing the main topologies of DC/AC converters intended for a wide range of applications. The proposed approach considers the reduction in the AC circuit in DC/AC converters to a series complex impedance with active-capacitive characteristics. This enables the switching of all semiconductor devices used in prototype power electronic systems, including single-operation thyristors. Despite the continuous development of modern semiconductor technology, thyristors remain relevant and economically advantageous for high-power applications.

The principal results and contributions of the research are as follows:

1. Based on the analysis of resonant DC/AC converters with and without reverse diodes, operating in various modes, the transfer functions of these devices have been derived. This makes it possible to unify the analysis of all operating modes, load-compensation circuit options, and DC/AC converters functioning in aperiodic mode (voltage source inverters).

2. A generalized analytical approach has been formulated to describe the behavior of resonant output circuits, relying on the phase shift angles between currents and voltages in the resonant elements.

3. The dependences of the U/U_{OUT} ratio for the most common configurations of complex output circuits in resonant DC/AC converters, operating under different modes, have been analytically obtained and graphically presented.

4. Design methodologies have been developed and validated for the main types of inverters with different application scopes.

By introducing suitable numerical coefficients, the proposed design methodologies can be adapted for other inverter configurations, including three-phase circuits and circuits with frequency doubling. The unified approach is suitable for formalization and algorithmic implementation, which enables its integration into specialized mathematical software for automated design processes.

An important direction for future development of this research is the application of artificial intelligence techniques to optimize and refine the inverter design process. By collecting and processing data from multiple design projects, it will be possible to create adaptive algorithms capable of improving accuracy, reducing design time, and enhancing the reliability and efficiency of power electronic devices and systems.

5. Conclusions

The present monograph provides a comprehensive and systematically structured study of DC/AC converters, which represent a fundamental class of devices in contemporary power electronics. The work encompasses the principal inverter types—current-source, voltage-source, and resonant—examining their operational principles, topological variations, control strategies and performance under diverse load and operating conditions.

A unified analytical and modeling framework has been developed, enabling the description, comparison, and optimization of various converter topologies within a common methodological basis. This approach not only facilitates high-accuracy prediction of electrical behavior but also supports the formalization of the design process, making it adaptable to different applications and scales. By integrating numerical simulations performed in MATLAB/Simulink and LTspice with analytical derivations, the study ensures a direct link between theoretical concepts and practical engineering solutions.

The proposed design methodology contributes to the standardization of engineering practice in the creation of modern inverter systems. Through representative case studies and verification by simulation, the applicability of the method is demonstrated in both conventional and advanced configurations, including those operating in resonant modes. The obtained deviations between calculated and simulated results remain within acceptable engineering limits, confirming the robustness of the methodology.

Looking ahead, several promising research directions emerge. The integration of model-based real-time optimization can enhance the adaptability of inverter systems to rapidly changing operating regimes. Furthermore, the application of artificial intelligence—particularly neural networks, genetic algorithms, and reinforcement learning—holds significant potential for developing intelligent control systems capable of self-optimization and predictive diagnostics. The concept of the digital twin is highlighted as a transformative tool for lifecycle management, enabling continuous monitoring, performance prediction, and proactive maintenance.

Ultimately, the findings of this monograph aim to serve both academia and industry by providing a solid theoretical foundation coupled with practical guidelines for the design and improvement of DC/AC converters. By fostering innovation in efficiency, reliability, and functionality, the work contributes to the advancement of power electronic technologies that underpin the sustainable and resilient energy systems of the future.

References

1. Mohan, N.; Undeland, T.M.; Robbins, W.P. *Power Electronics: Converters, Applications, and Design*, 3rd ed.; John Wiley & Sons, Inc.: Hoboken, NJ, USA, 2007.
2. Rashid, M.H. *Power Electronics Handbook*; Elsevier: Amsterdam, The Netherlands, 2023. [CrossRef]
3. Erickson, R.W.; Maksimović, D. *Fundamentals of Power Electronics*; Springer: Cham, Germany, 2020. [CrossRef]
4. Kassakian, J.G.; Perreault, D.J.; Verghese, G.C.; Schlecht, M.F. *Principles of Power Electronics*; Cambridge University Press: Cambridge, Germany, 2023. [CrossRef]
5. Zinoviev, G. *Fundamentals of Power Electronics-Part II*; Novosibirsk State University: Novosibirsk, Russia, 2004. (In Russian)
6. Pollefliet, J. *Power Electronics: Switches and Converters*; Elsevier: Amsterdam, The Netherlands, 2017. [CrossRef]
7. Hasan, M. Application of power electronics in power systems. In *Handbook of Research on Power and Energy System Optimization*; IGI Global Scientific Publishing: Hershey, PA, USA, 2018. [CrossRef]
8. da Silva, E.R.C.; Elbuluk, M.E. Fundamentals of Power Electronics. In *Green Energy and Technology*; Springer: London, UK, 2013; Volume 59. [CrossRef]
9. Rajabi, A.; Marangalu, M.G.; Shahir, F.M.; Sedaghati, R. Power electronics converters-An overview. In *Intelligent Control of Medium and High Power Converters*; The Institution of Engineering and Technology: Stevenage, UK, 2023. [CrossRef]
10. Robinson, F.V.P. Power electronics converters, applications and design. *Microelectron. J.* **1997**, *28*, 105–106. [CrossRef]
11. Amaya, L.E.; Krein, P.T.; Najm, F.N. A Design Methodology for Power Electronics. In Proceedings of the 5th IEEE Workshop on Computers in Power Electronics (COMPEL), Portland, OR, USA, 11–14 August 1996; pp. 52–57. [CrossRef]
12. Rashid, M.H. *Power Electronics—Circuits, Devices, and Applications*, 3rd ed.; Prentice Hall India: New Delhi, India, 2008; Volume 79.
13. Dokic, B.L.; Blanusa, B. *Power Electronics: Converters and Regulators*; Energy; Springer: Cham, Germany, 2014.
14. Haque, A.; Shah, N.; Malik, J.A.; Malik, A. Fundamentals of power electronics in smart cities. In *Smart Cities: Power Electronics, Renewable Energy, and Internet of Things*; CRC Press: Boca Raton, FL, USA, 2024. [CrossRef]
15. Popov, E. *Analysis, Modeling and Design of Converter Units (Computer–Aided Design of Power Electronic Circuits)*; Technical University Printing House, Sofia: Sofia, Bulgaria, 2005; pp. 59–80.
16. Todorov, T.S.; Madzharov, N.D.; Alexiev, D.T.; Ivanov, P.T. *Autonomous Inverters*; Technical University of Gabrovo: Gabrovo, Bulgaria, 1996.
17. Berkovich, E.I.; Ivenskyi, G.V. *High-Frequency Thyristor Converters for Electrical Devices*; Energoatomizdat: Leningrad, Russia, 1983.
18. Kazimierczuk, M.K.; Czarkowski, D. *Resonant Power Converters*; John Wiley & Sons: Hoboken, NJ, USA, 2012.

19. Batarseh, I.; Harb, A. *Power Electronics: Circuit Analysis and Design*; Springer: Cham, Germany, 2017. [CrossRef]
20. Antchev, M.H. *Technologies for Electrical Power Conversion, Efficiency, and Distribution: Methods and Processes*; IGI Global: Hershey, PA, USA, 2009. [CrossRef]
21. Skvarenina, T.L. *The Power Electronics Handbook: Industrial Electronics Series*; CRC Press: Boca Raton, FL, USA, 2001.
22. Tang, Z.; Yang, Y.; Blaabjerg, F. Power electronics: The enabling technology for renewable energy integration. *CSEE J. Power Energy Syst.* **2022**, *8*, 39–52. [CrossRef]
23. Zhang, G.; Li, Z.; Zhang, B.; Halang, W.A. Power electronics converters: Past, present and future. *Renew. Sustain. Energy Rev.* **2018**, *81*, 2028–2044. [CrossRef]
24. Ackermann, B.; Sauerländer, G. Power electronics for the light of the future. In *Bauelemente der Leistungselektronik und ihre Anwendungen 2017–7. ETG-Fachtagung*; VDE VERLAG GmbH: Berlin–Offenbach, Germany, 2017.
25. Rodriguez, J.; Blaabjerg, F.; Kazmierkowski, M.P. Energy Transition Technology: The Role of Power Electronics. *Proc. IEEE* **2023**, *111*, 329–334. [CrossRef]
26. Kumar, N.; Guerrero, J.M.; Kastha, D.; Saha, T.K. Power electronics for next-generation drives and energy systems. In *Clean Generation and Power Grids*; Institution of Engineering and Technology: London, UK, 2022; Volume 2.
27. Khluabwannarat, P.; Thammarat, C.; Tadsuan, S.; Bunjongjit, S. An analysis of iron loss supplied by sinusoidal, square wave, bipolar PWM inverter and unipolar PWM inverter. In Proceedings of the 8th International Power Engineering Conference, IPEC 2007, Singapore, 3–6 December 2007.
28. Bose, B.K. Power Electronics: My Life and Vision for the Future [My View]. *IEEE Ind. Electron. Mag.* **2022**, *16*, 65–72. [CrossRef]
29. Liang, T.J.P.; Blaabjerg, F.; Tan, D.; Ojo, O. Editorial Special Issue on Future of Power Electronics: Components, Circuits, and Systems. *IEEE J. Emerg. Sel. Top. Power Electron.* **2023**, *11*, 2399. [CrossRef]
30. Ninomiya, T.; Higashi, T.; Harada, K.; Nakahara, M. A Unified Analysis of Resonant Converters. *IEEE Trans. Power Electron.* **1991**, *6*, 260–270. [CrossRef]
31. Yu, S.Y. A new compact and high efficiency resonant converter. In Proceedings of the Conference Proceedings—IEEE Applied Power Electronics Conference and Exposition—APEC, Long Beach, CA, USA, 20–24 March 2016. [CrossRef]
32. Rudnev, V.; Loveless, D.; Cook, R.L. *Handbook of Induction Heating*, 2nd ed.; CRC Press: New York, NY, USA, 2017. [CrossRef]
33. Hinov, N. A Unified Approach to the Analysis of DC/AC Converters, Based on the Study of Electromagnetic Processes in a Series RLC Circuit. *Electronics* **2023**, *12*, 983. [CrossRef]
34. Hinov, N. New method for designing serial resonant power converters. *AIP Conf. Proc.* **2017**, *1910*, 60019.
35. Hinov, N.L. Power Converters of Electrical Energy with Industrial Application. Ph.D. Thesis, Technical University, Sofia, Bulgaria, 1998. (In Bulgarian).
36. Hinov, N. An Innovative Design Approach for Resonant DC/AC Converters, Based on Symmetry in Their Operating Modes. *Symmetry* **2023**, *15*, 1864. [CrossRef]

37. Gradinarov, N.P.; Hinov, N.L.; Marinov, T.L. General Analysis of Resonant Inverters with Reverse Diodes, Operating at Various Modes. In Proceedings of the 40th International Conference on Electrical Drives and Power Systems (ICEST), University of Niš, Faculty of Electronic Engineering, Niš, Serbia and Montenegro, 29 June–1 July 2005.
38. Lascu, D.; Lascu, M. Unified models for the series resonant converters. In Proceedings of the 6th International Conference on Optimization of Electrical and Electronic Equipments, OPTIM 1998, Braşov, Romania, 14–15 May 1998. [CrossRef]
39. Rico, M.; Garcia, V.; Hernando, M.; Nuno, F.; Sebastian, J.; Uceda, J. A unified method to model resonant switching converters. In Proceedings of the IECON Proceedings (Industrial Electronics Conference), Pacific Grove, CA, USA, 27–30 November 1990. [CrossRef]
40. Li, X.; Zhang, X.; Lin, F.; Blaabjerg, F. Artificial-Intelligence-Based Design for Circuit Parameters of Power Converters. *IEEE Trans. Ind. Electron.* **2022**, *69*, 11144–11155. [CrossRef]
41. Qashqai, P.; Vahedi, H.; Al-Haddad, K. Applications of artifical intelligence in power electronics. In Proceedings of the 2019 IEEE 28th International Symposium on Industrial Electronics (ISIE), Vancouver, BC, Canada, 12–14 June 2019. [CrossRef]
42. Tian, F.; Cobaleda, D.B.; Martinez, W. Artificial-Intelligence based DC-DC Converter Efficiency Modelling and Parameters Optimization. In Proceedings of the 24th European Conference on Power Electronics and Applications, EPE 2022 ECCE Europe, Hannover, Germany, 5–9 September 2022.
43. Padmanaban, S.; Nasab, M.A.; Samavat, T.; Zand, M.; Nasab, M.A. Artificial Intelligence Techniques for Smart Power Systems. In *IoT and Analytics in Renewable Energy Systems (Volume 1): Sustainable Smart Grids and Renewable Energy Systems*; CRC Press: Boca Raton, FL, USA, 2023. [CrossRef]
44. Madzharov, N. High-frequency power source with constant output power. *J. Eng. Sci. Technol. Rev.* **2016**, *9*, 157–162. [CrossRef]
45. Hinov, N.; Arnaudov, D.; Madzharov, N. Comparison of converters with constant output power for wireless power transmission. In Proceedings of the PCIM Europe 2019; International Exhibition and Conference for Power Electronics, Intelligent Motion, Renewable Energy and Energy Management, Nuremberg, Germany, 7–9 May 2019; pp. 1–8.
46. Madzharov, N.; Iliev, D. Converters with Energy Dosing by Inductance. In Proceedings of the 2023 58th International Scientific Conference on Information, Communication and Energy Systems and Technologies, ICEST 2023, Nis, Serbia, 29 June–1 July 2023. [CrossRef]
47. Madzharov, N.; Prodanov, P. Reliability Analysis of Autonomous Inverter with Energy Dosing for Various Applications. *AIP Conf. Proc.* **2022**, *2505*, 080022. [CrossRef]
48. Holmes, D.G.; Lipo, T.A. Introduction to Power Electronic Converters. In *Pulse Width Modulation for Power Converters*; Wiley-IEEE Press: Piscataway, NJ, USA, 2010. [CrossRef]
49. Pollefliet, J. Inverters. In *Power Electronics*; Elsevier Inc.: Amsterdam, The Netherlands, 2018. [CrossRef]

50. Agarwal, V. DC-AC converters (inverters). In *Encyclopedia of Electrical and Electronic Power Engineering: Volumes 1–3*; Elsevier Inc.: Amsterdam, The Netherlands, 2022; Volume 3. [CrossRef]
51. Luo, F.L.; Ye, H. Current-source inverters. In *Advanced DC/AC Inverters*; CRC Press: Boca Raton, FL, USA, 2017. [CrossRef]
52. Gao, D.Z.; Sun, K. DC-AC inverters. In *Electric Renewable Energy Systems*; Elsevier Inc.: Amsterdam, The Netherlands, 2016. [CrossRef]
53. Tleis, N. Modelling of voltage-source inverters, wind turbine and solar photovoltaic (PV) generators. In *Power Systems Modelling and Fault Analysis*; Elsevier Inc.: Amsterdam, The Netherlands, 2019. [CrossRef]
54. Gao, B.; Wang, Y.; Xu, W. Modeling Voltage Source Converters for Harmonic Power Flow Studies. *IEEE Trans. Power Deliv.* **2021**, *36*, 3426–3437. [CrossRef]
55. Hinov, N. New Method for Analysis and Design Consideration of Voltage Source Inverters. *Energies* **2022**, *15*, 3695. [CrossRef]
56. Hsieh, Y.H.; Lee, F.C. Design-Oriented Equivalent Circuit Model for Resonant Converters. In Proceedings of the IEEE Applied Power Electronics Conference and Exposition—APEC, New Orleans, LA, USA, 15–19 March 2020. [CrossRef]
57. Penev, D.; Arnaudov, D.; Hinov, N. Formalization, equivalence and generalization of basic resonance electrical circuits. *AIP Conf. Proc.* **2017**, *1910*, 60025. [CrossRef]
58. Padmanaban, S.; Palanisamy, S.; Chenniappan, S.; Holm-Nielsen, J.B. *Artificial Intelligence-Based Smart Power Systems*; Wiley: Piscataway, NJ, USA, 2022. [CrossRef]
59. Zhang, Z.; Wen, F.; Sun, Z.; Guo, X.; He, T.; Lee, C. Artificial Intelligence-Enabled Sensing Technologies in the 5G/Internet of Things Era: From Virtual Reality/Augmented Reality to the Digital Twin. *Adv. Intell. Syst.* **2022**, *4*, 2100228. [CrossRef]
60. Wolf, K.H. Artificial Intelligence (AI) and Expert Systems (ES). *Ore Geol. Rev.* **1994**, *9*, 241–243. [CrossRef]
61. Hinov, N.L. Model-Based Design of Power Electronic Devices. Ph.D. Thesis, Technical University, Sofia, Bulgaria, 2024. (In Bulgarian).
62. Combe, Q.; Pierfederici, S.; Weber, M.; Dufour, S. Modeling, Analysis and Control of Current Source Converter. In Proceedings of the IECON Proceedings (Industrial Electronics Conference), Toronto, ON, Canada, 13–16 October 2021. [CrossRef]
63. Lin, Y.-L. Resonant DC/DC Converters: Modeling, Control Strategies, Simulation, and Experimental Studies. Master's Thesis, Rochester Institute of Technology, Rochester, NY, USA, 1986.
64. Qiao, Z.; Yang, R.; Liao, W.Q.; Yu, W.; Fang, Y. Impedance modeling, Parameters sensitivity and Stability analysis of hybrid DC ship microgrid. *Electr. Power Syst. Res.* **2024**, *226*, 109901. [CrossRef]
65. Mehrabadi, N.R.; Burgos, R.; Roy, C.; Boroyevich, D. Power electronics modeling and design. *IEEE Power Electron. Mag.* **2017**, *4*, 44–52. [CrossRef]
66. Bacha, S.; Li, H.; Montenegro-Martinez, D. Complex Power Electronics Systems Modeling and Analysis. *IEEE Trans. Ind. Electron.* **2019**, *66*, 6412–6415. [CrossRef]

67. Gole, A.M.; Keri, A.; Kwankpa, C.; Gunther, E.W.; Dommel, H.W.; Hassan, I.; Marti, J.R.; Martinez, J.A.; Fehrle, K.G.; Tang, L.; et al. Guidelines for modeling power electronics in electric power engineering applications. *IEEE Trans. Power Deliv.* **1997**, *12*, 505–514. [CrossRef]
68. Maksimovic, D.; Stankovic, A.M.; Thottuvelil, V.J.; Verghese, G.C. Modeling and Simulation of Power Electronic Converters. *Proc. IEEE* **2001**, *89*, 898–912. [CrossRef]
69. Senesky, M.; Eirea, G.; Koo, T.J. *Hybrid Modelling and Control of Power Electronics*; Lecture Notes in Computer Science (Including Subseries Lecture Notes in Artificial Intelligence and Lecture Notes in Bioinformatics); Springer: Berlin/Heidelberg, Germany, 2003; Volume 2623. [CrossRef]
70. Simões, M.G.; Farret, F.A. *Modeling Power Electronics and Interfacing Energy Conversion Systems*; John Wiley & Sons: Hoboken, NJ, USA, 2016. [CrossRef]
71. Asadi, F. *Simulation of Power Electronics Circuits with MATLAB®/Simulink®: Design, Analyze, and Prototype Power Electronics*; Apress: New York, NY, USA, 2022. [CrossRef]
72. Raghuwanshi, S.S.; Singh, A.; Mokhariwale, Y. A Comparison & Performance of Simulation Tools MATLAB/SIMULINK, PSIM & PSPICE for Power Electronics Circuits. *Int. J. Adv. Res. Comput. Sci. Softw. Eng.* **2012**, *2*, 187–191.
73. Gu, D.W.; Petkov, P.H.; Konstantinov, M.M. Robust control design with MATLAB®. In *Advanced Textbooks in Control and Signal Processing*; Elsevier Inc.: Amsterdam, The Netherlands, 2013. [CrossRef]
74. Asadi, F. Simulation of Power Electronics Circuits with LTspice®. In *Essential Circuit Analysis Using LTspice®*; Springer: Cham, Germany, 2023. [CrossRef]
75. Krishnamoorthy, H.S.; Krein, P.; Zahnstecher, B. From 'Power Electronics Inside' to 'Human-Centered Power Electronics'. *IEEE Power Electron. Mag.* **2023**, *10*, 61–63. [CrossRef]
76. Cerezo, J. IGBT Definition for Single Ended Induction Heating Cookers. In *Bodo's Power Systems: Electronics in Motion and Conversion*; A Media Print: Laboe, Germany, 12 April 2012; pp. 22–30. Available online: https://www.bodospower.com/restricted/downloads/bp_2012_04.pdf (accessed on 2 October 2024).
77. Dimitrov, B.; Hayatleh, K.; Barker, S.; Collier, G. Design, Analysis and Experimental Verification of the Self-Resonant Inverter for Induction Heating Crucible Melting Furnace Based on IGBTs Connected in Parallel. *Electricity* **2021**, *2*, 439–458. [CrossRef]
78. Freeland, S.; Middlebrook, R.D. Unified Analysis of Converters with Resonant Switches. In Proceedings of the PESC Record—IEEE Annual Power Electronics Specialists Conference, Blacksburg, VA, USA, 21–26 June 1987. [CrossRef]
79. Li, X.; Chen, S.; Tang, Y.; Zhang, Y.; Zhang, X. Reduced-Order Equivalent Circuit Model of Series Resonant Converter Considering the Interaction between Resonant Elements. In Proceedings of the 2021 IEEE Energy Conversion Congress and Exposition, ECCE 2021—Proceedings, Vancouver, BC, Canada, 10–14 October 2021. [CrossRef]
80. Xu, X.; Li, Y.; Niu, R.; Tian, L.; Li, Y.; Liang, B. Enhanced Robust Control of the DC-DC Converters Based on Reduced-Order GPI Observer. In Proceedings of the 2023 26th International Conference on Electrical Machines and Systems, ICEMS 2023, Zhuhai, China, 5–8 November 2023. [CrossRef]

81. Khan, H.; Bazaz, M.A.; Nahvi, S.A. Model order reduction of power electronic circuits. In Proceedings of the 2017 6th International Conference on Computer Applications in Electrical Engineering—Recent Advances, CERA 2017, Roorkee, India, 5–7 October 2017. [CrossRef]
82. Dolinina, A. Macromodeling of electronic circuits based on model order reduction. In Proceedings of the 2015 IEEE North West Russia Section Young Researchers in Electrical and Electronic Engineering Conference, ElConRusNW 2015, St. Petersburg, Russia, 2–4 February 2015. [CrossRef]
83. Roberts, P.M. *Introduction to Brazing Technology*; CRC Press: Boca Raton, FL, USA, 2016. [CrossRef]
84. Hinov, N. Design consideration of resonance inverters with electro-technological application. *AIP Conf. Proc.* **2017**, *1910*, 060024. [CrossRef]
85. Hinov, N. Quasi-Boundary Method for Design Consideration of Resonant DC-DC Converters. *Energies* **2021**, *14*, 6153. [CrossRef]
86. Hinov, N. New method for designing parallel loaded DC-AC power converters. *AIP Conf. Proc.* **2018**, *2048*, 60022. [CrossRef]

Index

MDPI AG
Grosspeteranlage 5
4052 Basel
Switzerland
Tel.: +41 61 683 77 34

MDPI Books Editorial Office
E-mail: books@mdpi.com
www.mdpi.com/books

www.ingramcontent.com/pod-product-compliance
Lightning Source LLC
Chambersburg PA
CBHW040852230926
51850CB00012BA/2934
* 9 7 8 3 7 2 5 8 4 0 5 8 8 *